Introduction to

PERMACULTURE

Introduction to
PERMACULTURE

Bill Mollison

with
Reny Mia Slay

TAGARI PUBLICATIONS TYALGUM AUSTRALIA

A TAGARI PUBLICATION

© Bill Mollison 1991
2nd edition 1994 7,000 copies
Reprinted 1994 5,000 copies
Reprinted 1995 6,000 copies
Reprinted 1996 3,000 copies
Reprinted 1997 5,000 copies

Editing/additional text: Marlyn Wade & Andrew Murray
Illustrations: Andrew Jeeves
Additional Illustrations: Janet Mollison, Kate Feain, Giri Mazzella, Claire Yerbury, Catherine Worsley
Cover: Kate Feain with Wayne Fleming
Photography: Craig Worsley

National Library of Australia
Cataloguing-in-Publication

Mollison, Bill
    Introduction to permaculture

    Bibliography.
    Includes index.
    ISBN 0 908228 08 2

    1. Permaculture.
    I. Slay, Reny. II. Title

631.58

Printed by Australian Print Group,
Maryborough, Victoria.

Tagari Publications
PO Box 1, Tyalgum, NSW 2484
Australia
Ph: (066) 793 442
Fax: (066) 793 567
Email: permacultureinstitute@compuserve.com

Table of Contents

Preface

I grew up in a small village in Tasmania. Everything that we needed we made. We made our own boots, our own metal works; we caught fish, grew food, made bread. I didn't know anybody who lived there who had only one job, or even anything that you could define as a job. Everybody worked at several things.

Until I was about 28, I lived in a sort of dream. I spent most of my time in the bush or on the sea. I fished, I hunted for my living. It wasn't until the 1950s that I noticed large parts of the system in which I lived were disappearing. Fish stocks started to collapse. Seaweed around the shorelines had thinned out. Large patches of the forest began to die. I hadn't realised until then that I had become very fond of them, that I was in love with my country.

After many years as a scientist with the CSIRO Wildlife Survey Section and with the Tasmanian Inland Fisheries Department, I began to protest against the political and industrial systems I saw were killing us and the world around us. But I soon decided that it was no good persisting with opposition that in the end achieved nothing. I withdrew from society for two years; I did not want to oppose anything ever again and waste my time. I wanted to come back only with something very positive, something that would allow us all to exist without the wholesale collapse of biological systems.

In 1968 I began teaching at the University of Tasmania, and in 1974, David Holmgren and I jointly evolved a framework for a sustainable agricultural system based on a multi-crop of perennial trees, shrubs, herbs (vegetables and weeds), fungi, and root systems, for which I coined the word "permaculture". We spent a lot of time working out the principles of permaculture and building a species-rich garden. This culminated, in 1978, in the publication of *Permaculture One*, followed a year later by *Permaculture Two*.

Public reaction to permaculture was mixed. The professional community was outraged, because we were combining architecture with biology, agriculture with forestry, and forestry with animal husbandry, so that almost everybody who considered themselves to be a specialist felt a bit offended. But the popular response was very different. Many people had been thinking along the same lines. They were dissatisfied with agriculture as it is now practised, and were looking towards more natural, ecological systems.

As I saw permaculture in the 1970s, it was a beneficial assembly of plants and animals in relation to human settlements, mostly aimed towards household and community self-reliance, and perhaps as a "commercial endeavour" only arising from a surplus from that system.

However, permaculture has come to mean more than just food-sufficiency in the household. Self-reliance in food is meaningless unless people have access to land, information, and financial resources. So in recent years it has come to encompass appropriate legal and financial strategies, including strategies for land access, business structures, and regional self-financing. This way it is a whole human system.

By 1976, I was lecturing on permaculture, and in 1979 I resigned from my teaching position and threw myself at an advanced age into an uncertain future. I decided to do nothing else but to try to persuade people to build good biological systems. I designed quite a few properties, and existed for a while by catching fish and pulling potatoes. In 1981 the first graduates of a standard permaculture design course also started to design permaculture systems in Australia.

Today there are over 12,000 such graduates throughout the world, all involved in some aspect of environmental and social work.

Bill Mollison

ACKNOWLEDGMENTS

We are grateful to the large number of students and practicing permaculturalists from all over the world who over the years have experimented with plant species, designed properties, written informative articles, set up permaculture organisations in their own states and countries, taught other students, and who have all helped make part of the earth a better place to live, not only for our children, but for us *now*.

ACCESS TO INFORMATION

Material in this book is accessed through the chapter and section contents (Table of Contents). Main subjects are listed in the Index. A list of the common and Latin names of plants used in this book and a glossary of a few uncommonly-used words are located at the back of the book, along with the Appendices, containing an extensive plant species listing and a directory of permaculture addresses and resources.

TREE TITHE

Each volume of *Introduction to Permaculture* carries a surcharge of 50¢ which is paid by Tagari Publications to the Permaculture Institute. The Institute holds these funds in trust for tree-planting, and from time to time releases monies to selected groups who are active in permanent reafforestation. In this way, both publishers and readers can have a clear conscience about the use of the paper in this volume, or in any book published by Tagari Publications.

CONVENTIONS USED

Seasons and directions: So that the text and figures are useful and readable in both the north and south hemispheres, the words "sun-side" or "sunwards" or "shade-side" or "polewards" are used rather than north and south. The symbol below is used in the illustrations to indicate the sun direction.

Introduction

Permaculture is a design system for creating sustainable human environments. The word itself is a contraction not only of **permanent agriculture** but also of permanent culture, as cultures cannot survive for long without a sustainable agricultural base and landuse ethic. On one level, permaculture deals with plants, animals, buildings, and infrastructures (water, energy, communications). However, permaculture is not about these elements themselves, but rather about the relationships we can create between them by the way we place them in the landscape.

The aim is to create systems that are ecologically-sound and economically viable, which provide for their own needs, do not exploit or pollute, and are therefore sustainable in the long term. Permaculture uses the inherent qualities of plants and animals combined with the natural characteristics of landscapes and structures to produce a life-supporting system for city and country, using the smallest practical area.

Permaculture is based on the observation of natural systems, the wisdom contained in traditional farming systems, and modern scientific and technological knowledge. Although based on good ecological models, permaculture creates a *cultivated* ecology, which is designed to produce more human and animal food than is generally found in nature.

Fukuoka, in his book *The One Straw Revolution*, has perhaps best stated the basic philosophy of permaculture. In brief, it is a philosophy of working with, rather than against nature; of protracted and thoughtful observation rather than protracted and thoughtless labour; and of looking at plants and animals in all their functions, rather than treating elements as a single-product system. I have spoken, on a more mundane level, of using aikido on the landscape, of rolling with the blows, turning adversity into strength, and using everything positively. The other approach is to karate the landscape, to try to make it yield by using our strength, and striking many hard blows. But if we attack nature we attack (and ultimately destroy) ourselves.

I think harmony with nature is possible only if we abandon the idea of superiority over the natural world. Levi Strauss said that our profound error is that we have always looked upon ourselves as "masters of creation", in the sense of being above it. We are not superior to other life-forms; all living things are an expression of Life. If we could see that truth, we would see that everything we do to other life-forms we also do to ourselves. A culture which understands this does not, without absolute necessity, destroy any living thing.

Permaculture is a system by which we can exist on the earth by using energy that is naturally in flux and relatively harmless, and by using food and natural resources that are abundant in such a way that we don't continually destroy life on earth. Every technique for conserving and restoring the earth is already known; what is not evident is that any nation or large group of people is prepared to make the change. However, millions of ordinary people are starting to do it themselves without help from political authorities.

1

Wherever we live, we should start to do something. We can start first by decreasing our energy consumption—you can actually live on 40% of the energy you are now using without sacrificing anything of value. We can re-fit our houses for energy efficiency. We can cut our vehicle use by using public transportation and sharing with friends. We can save water off our roofs into tanks, or recycle greywater to the toilet system or garden. We can also begin to take some part in food production. This doesn't mean that we all need to grow our own potatoes, but it may mean that we will buy them directly from a person who is already growing potatoes responsibly. In fact, one would probably do better to organise a farmer-purchasing group in the neighbourhood than to grow potatoes.

is regarded as a commodity involves a shift from a low to a high-energy society, the use of land in an exploitative and destructive way, and a demand for external energy sources, mainly provided by the third world as fuels, fertilisers, protein, labour, and skills.

Conventional farming does not recognise and pay its true costs: the land is mined of its fertility to produce annual grain and vegetable crops; non-renewable resources are used to support yields; the land is eroded through over-stocking of animals and extensive ploughing; land and water are polluted with chemicals.

When the needs of a system are not met from within the system, we pay the price in energy consumption and pollution. We can no longer afford the true cost of our agriculture. It is

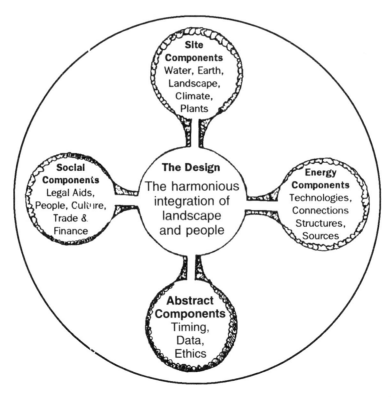

ELEMENTS OF A TOTAL PERMACULTURE DESIGN

In all permanent agricultures, or in sustainable human culture generally, the energy needs of the system are provided by that system. Modern crop agriculture is totally dependent on external energies. The shift from productive permanent systems (where the land is held in common), to annual, commercial agricultures where land

killing our world, and it will kill us.

Sitting at our back doorsteps, all we need to live a good life lies about us. Sun, wind, people, buildings, stones, sea, birds and plants surround us. Cooperation with all these things brings harmony, opposition to them brings disaster and chaos.

PERMACULTURE ETHICS

Ethics are moral beliefs and actions in relation to survival on our planet. In permaculture, we embrace a threefold ethic: care of the earth, care of people, and dispersal of surplus time, money, and materials towards these ends.

Care of the earth means care of all living and nonliving things: soils, species and their varieties, atmosphere, forests, micro-habitats, animals, and waters. It implies harmless and rehabilitative activities, active conservation, ethical and frugal use of resources, and "right livelihood" (working for useful and beneficial systems).

Care of the earth also implies *care of people* so that our basic needs for food, shelter, education, satisfying employment, and convivial human contact are taken care of. Care of people is important, for even though people make up a small part of the total living systems of the world, we make a decisive impact on it. If we can provide for our basic needs, we need not indulge in broadscale destructive practices against the earth.

The third component of the basic "care of the earth" ethic is the *contribution of surplus time, money, and energy* to achieve the aims of earth and people care. This means that after we have taken care of our basic needs and designed our systems to the best of our ability, we can extend our influence and energies to helping others achieve that aim.

The permaculture system also has a basic *life ethic*, which recognises the intrinsic worth of every living thing. A tree is something of value *in itself*, even if it has no commercial value for us. That it is alive and functioning is what is important. It is doing its part in nature: recycling biomass, providing oxygen and carbon dioxide for the region, sheltering small animals, building soils, and so on.

So we see that the permaculture ethic pervades all aspects of environmental, community, economic and social systems. *Cooperation, not competition, is the key.*

Ways we can implement the earthcare ethics in our own lives are as follows:

• Think about the long-term consequences of your actions. Plan for sustainability.

• Where possible use species native to the area, or those naturalised species known to be beneficial. The thoughtless introduction of potentially invasive species may upset natural balances in your home area.

• Cultivate the smallest possible land area. Plan for small-scale, energy-efficient *intensive* systems rather than large-scale, energy-consuming *extensive* systems.

• Be diverse, polycultural (as opposed to monocultural). This provides stability and helps us to be ready for change, whether environmental or social.

• Increase the sum of yields: look at the *total* yield of the system provided by annuals, perennials, crops, trees, and animals. Also regard energy saved as a yield.

• Use low-energy environmental (solar, wind, and water) and biological (plant and animal) systems to conserve and generate energy.

• Bring food-growing back into the cities and towns, where it has always traditionally been in sustainable societies.

• Assist people to become self-reliant, and promote community responsibility.

• Reafforest the earth and restore fertility to the soil.

• Use everything at its optimum level and recycle all wastes.

• See solutions, not problems.

• Work where it counts (plant a tree where it will survive; assist people who want to learn).

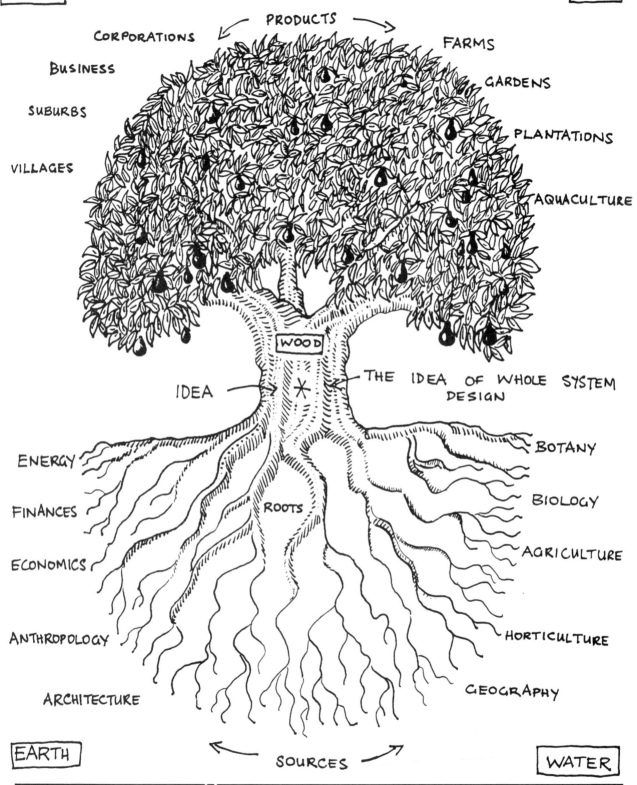

THE PERMACULTURE TREE
The Permaculture Tree, the elements of design. Roots are in many disciplines, an abstract world. Products lie in the real world. The germination of an idea translates in formation into product. (The five elements; wood, fire (light), earth, air, water, are organised by a tree, as information is organised by ideas).

CHAPTER 1

Permaculture Principles

1.1 INTRODUCTION

There are two basic steps to good permaculture design. The first deals with laws and principles that can be adapted to any climatic and cultural condition, while the second is more closely associated with practical techniques, which change from one climate and culture to another.

The principles discussed in the following pages are inherent in any permaculture design, in any climate and at any scale. They are selected from principles of various disciplines: ecology, energy conservation, landscape design, and environmental science, and are, briefly, as follows:

• Relative location: every element (such as house, pond, road, etc.) is placed in relationship to another so that they assist each other.

• Each element performs many functions.

• Each important function is supported by many elements.

• Efficient energy planning for house and settlement (zones and sectors).

• Emphasis on the use of biological resources over fossil fuel resources.

• Energy recycling on site (both fuel and human energy).

• Using and accelerating natural plant succession to establish favourable sites and soils.

• Polyculture and diversity of beneficial species for a productive, interactive system.

• Use of edge and natural patterns for best effect.

1.2 RELATIVE LOCATION

The core of permaculture is design. Design is a connection between things. It's not water, or a chicken, or the tree. It is how the water, the chicken and the tree are connected. It's the very opposite of what we are taught in school. Education takes everything and pulls it apart and makes no connections at all. Permaculture makes the connection, because as soon as you've got the connection you can feed the chicken from the tree. To enable a design component (pond, house, woodlot, garden, windbreak, etc.) to function efficiently, *we must put it in the right place.*

For example, dams and water tanks are located above the house and garden so that gravity rather than a pump is used to direct flow. Home windbreaks are placed so that they deflect wind but do not shade the house from the winter sun. The garden is placed between the house and the chicken pen, so that garden refuse is collected on the way to the pen and chicken manure is easily shovelled over to the garden, and so on.

We set up working relationships between each element, so that the needs of one element

5

are filled by the yields of another element. To do this, we must discover the basic characteristics of any element, its needs, and its products (see **Box**).

The elements in a typical small farm might include: house, greenhouse, garden, chicken pens, water storage tank, compost pile, beehives, nursery area and potting shed, woodlot, dam, aquaculture pond, windbreak, barn, tool shed, woodpile, guest-house, pasture, hedgerow, worm beds, and so on. These can be moved about, on paper, until they are working to best advantage.

In the case of every element, we can base our linking strategies to these questions:

"Of what use are the products of this particular element to the needs of other elements?"

"What needs of this element are supplied by other elements?"

"Where is this element incompatible with other elements?"

"Where does this element benefit other parts of the system?"

It is best to start with the most important node of activity (e.g. the house, or even a commercial centre such as nursery, free-range chicken farm, aquaculture, etc.). For things to work properly, we must remember that:

• The inputs needed by one element are supplied by other elements in the system; and

• The outputs needed by one element are used by other elements (including ourselves).

1.3
EACH ELEMENT PERFORMS MANY FUNCTIONS

Each element in the system should be chosen and placed so that it performs as many functions as possible. A pond can be used for irrigation, watering livestock, aquatic crop, and fire control. It is also a habitat for waterfowl, a fish farm, and a light reflector (**Figure 2.8**). A dam wall functions as a road, a firebreak, and a bamboo production area.

We can do the same with plants. Simply by selecting a useful species and putting it in a particular place we can use it for one or more of the following purposes:

windbreak	animal forage
privacy	fuel
trellis	erosion control
fire control	wildlife habitat
mulch	climatic buffer
food	soil conditioner

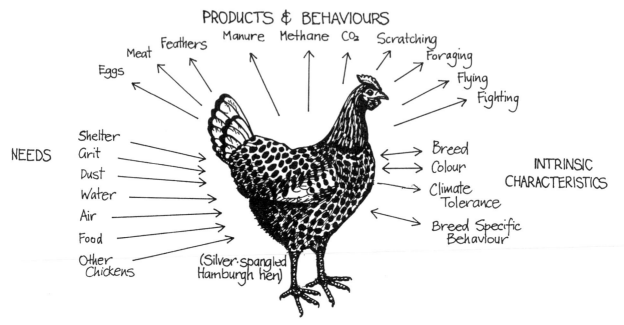

FIGURE 1.1 Analysing the characteristics, needs, and products of each element in the system in order to put it in the right place relative to other elements in the system.

Functional Analysis of the Chicken

This is how the process of relative location works, using the chicken as an example.

First, we list the innate characteristics of the chicken: its colour, size and weight, heat and cold tolerances, ability to rear its own young, etc. Chickens have different breed characteristics: light-coloured chickens tolerate heat better than dark-coloured ones; heavy breeds cannot fly as high as light breeds (which means fencing height requirements are different); some breeds are better mothers, others are better layers. We also look at the behaviour of the chicken: what is its "personality"? We see that all chickens scratch for food, walk, fly, roost in trees or perches at night, form flocks, and lay eggs.

Secondly we list basic needs:

Chickens need shelter, water, a dust bath to deter lice, a protected roosting area, and nest boxes. They need a source of shell grit to grind food around in their crops. And they like to be with other chickens. A solitary chicken is a pretty sad affair—best to give it a few companions. That's all easy enough to provide and wouldn't take us more than a few days to set up. Chickens also need food, and that's where we start to make connections to the other elements in our system, because we want to put the chicken in a place and situation where it will scratch for its own living. Any time we stop the chicken from behaving naturally—i.e. foraging—we've got to do the work for it. Both work and pollution are the result of incorrectly designed or unnatural systems.

Lastly, we list the products or outputs of the chicken. It provides meat, eggs, feathers, feather dust, manure, carbon dioxide (from breathing), sound, heat, and methane. We will want to place the chicken in such a position that its products are used by other elements in the system. Unless we use these outputs to aid some other part of our system, we are faced with more work and pollution.

Now we have all the information needed to sketch a plan of the chicken run, to decide where fences, shelters, nests, trees, seed and green crops, ponds, greenhouses, and processing centres will go *relative to the chicken*. Thus,

The House needs food, cooking fuel, heat in cold weather, hot water, lights, etc. It gives shelter and warmth for people. The chicken can supply some of these needs (food, feathers, meth-ane). It also consumes most food wastes coming from the house.

The Garden needs fertiliser, mulch, water. It gives leaves, seeds, vegetables. The chicken provides manures and eats surplus garden products. Chicken-pens close to the garden ensure easy collection of manures and a throw-over-the-fence feeding system. Chickens can be let into the garden, but only under controlled circumstances.

The Greenhouse needs carbon dioxide for plants, methane for germination, manure, heat, and water. It gives heat by day, and food for people, with some crop wastes for chickens. The chicken can obviously supply many of these needs, and utilise most of the wastes. It can also supply night heat to the greenhouse in the form of body heat if we place the chicken-house adjoining it (**Figure 7.8**).

The Orchard needs weeding, pest control, manure, and some pruning. It gives food (fruit and nuts), and provides insects for chicken forage. Thus, the orchard and the chicken can interact beneficially if chickens are allowed in from time to time.

The Woodlot needs management, fire control, perhaps pest control, some manure. It gives solid fuel, berries, seeds, insects, shelter, and some warmth. Chickens can roost in the trees, feed upon insect larvae, and assist in fire control by scratching or grazing fuels such as grasses.

The Cropland needs ploughing, manuring, seeding, harvesting, and storage of crop. It gives food for chickens and people. Chickens have a part to play as manure providers and cultivators (a large number of chickens on a small area will effectively clear all vegetation and turn the soil over by scratching).

The Pasture needs cropping, manuring, and storage of hay or silage. It gives food for animals (worms and insects included).

The Pond needs some manure. It yields fish, water plants as food, and can reflect light and absorb heat.

Simply by letting chickens behave naturally and range where they are of benefit, we get a lot of "work" out of them. Using the information above, we place the chicken near the (fenced) garden, and probably backing onto the greenhouse. Gates are opened at appropriate times into the orchard, pasture, and woodlot so that chickens forage fallen fruit, seeds, and insects, scratching out weeds and leaving behind manures.

A windbreak can be made up of trees that provide fodder or sugar pods for cows (willow, honey locust, tagasaste, taupata, carob); coppice for kindling or firewood (*Leucaena*); give nectar and pollen for bees (*Acacia fimbriata*); and provide for their own nitrogen requirements (leguminous trees). Acacias fulfil many functions: they provide seeds for poultry forage, foliage for larger stock, and fix nitrogen in the soil, while blossoms provide pollen for bees. They are also pioneer plants which prepare and protect the soil for slower-growing, more sensitive plants.

Selecting appropriate species requires a thorough knowledge of the animal or plant cultivar under consideration, its tolerances, its needs and its products. When considering plants, for example, we want to know: Is it deciduous or evergreen? Are its roots invasive? To what height does it grow? Is it fast-growing and short-lived, or slow-growing and long-lived? Does it have a dense or light canopy? Is it disease-resistant, or susceptible? Can it be browsed or cut, or will it die if over-pruned or coppiced?

To begin, start a species index, or keep notes on each plant (its characteristics, tolerances, and uses) on cards in a file system (see the annotated species lists in the Appendix). Some of the things to note are as follows:

1. **Form**: life style (annual, perennial, deciduous, evergreen) and shape (shrub, vine, tree), including heights.

2. **Tolerances**: climatic zone (arid, temperate, tropical, subtropical); shade or sun tolerance (preferring shade, partial shade, full sun); habitat (moist, dry, wet, high or low elevation); soil tolerance (sandy, clayey, rocky); and pH tolerance (acidic or alkaline soils).

3. **Uses**: edible (human food or seasoning); medicinal; animal forage (for specific animals, e.g. chickens, pigs, deer); soil improvement (nitrogen fixing, cover crop and green manure); site protection (erosion control, living fence,

windbreak); coppicing (fuel, poles, stakes); building material (poles, timber, furniture); and other uses (fibre, fuel, insect control, ornamental, nectar and pollen for bees, rootstock, dye).

There are various factors that may limit species selection:

• Unsuitable for climate or soil.

• Locally rampant or noxious.

• Unavailable or rare (usually not traded outside the country of origin).

• Preference (vegetarians may not choose fodder species or animals used for meat).

• Area of land available (smaller species for small properties).

• Usefulness in relation to difficulty of growing, small yield, or time taken to reach maturity.

1.4
EACH IMPORTANT FUNCTION IS SUPPORTED BY MANY ELEMENTS

Important basic needs such as water, food, energy, and fire protection should be served in two or more ways. A careful farm design, for example, will include both annual and perennial pasture *and* fodder trees (poplars, willows, honey locust, and tagasaste) which are either cut and fed to domestic stock, or the stock let in for short periods of time to eat the leaves, pods, or lopped branches.

In the same way, a house with a solar hot water system may also contain a back-up wood-burning stove with a water jacket to supply hot water when the sun is not shining. And for fire control, many elements (the pond, driveway, slow-burning windbreak trees, and swales) are incorporated in the homestead or village design to reduce damage should wildfire occur.

In other examples, water is caught in a variety of ways, from dams and tanks to swales and chisel ploughing (to replenish ground waters), and on sea coasts, winds are contained first by a strong, frost-line windbreak of trees and shrubs, and closer in by semi-permeable fences or trellis systems.

The key to efficient energy planning (which is, in fact, efficient economic planning) is the zone and sector placement of plants, animal ranges, and structures. The only modifiers are local factors of market, access, slope, local climatic quirks, areas of special interest (flood plains or rocky hillsides), and special soil conditions, such as hard laterites or swamp soils. The following sections cover zone, sector, and slope plans for an "ideal" site, say a gentle slope facing the sun side where few variables are encountered. "Real" landscape, however, will differ, so that your designs will be more complex than those illustrated.

■ ZONE PLANNING

Zone planning means placing elements according to how much we use them or how often we need to service them. Areas that must be visited every day (e.g. the glasshouse, chicken pen, garden) are located nearby, while places visited less frequently (orchard, grazing areas, woodlot) are located further away (**Figure 1.2**). To place elements in zones, start from a centre of activity, usually the house, although this can also be a barn, a plant nursery business, or, on a larger scale, an entire village.

Zoning is decided by (1) the number of times you need to visit the element (plant, animal or structure) for harvest or yield; and (2) the number of times the element needs you to visit it.

For example, on a yearly basis, we might visit the poultry shed:
- for eggs, 350 times;
- for manure, 20 times;
- for culling, 5 times;
- other, 20 times.

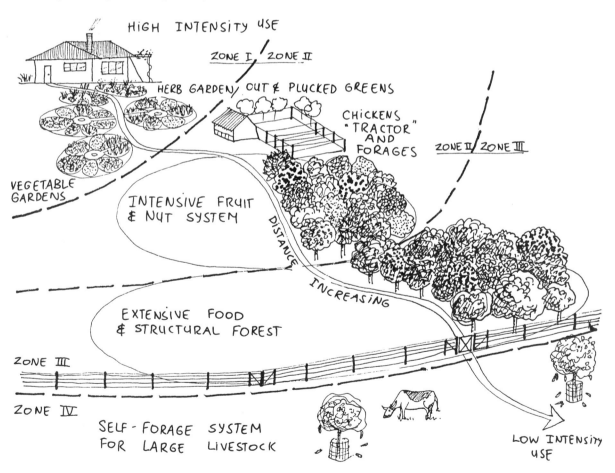

FIGURE 1.2 The relationship between distance and intensity of use. Frequently-visited areas are placed closest to the house.

For a total of 395 visits annually; whereas one might visit an oak tree only twice a year, to collect acorns. The more visits needed, the closer the objects need to be. Those components needing very frequent observation, constant visits, work input, or complex management techniques must be placed very close by, or we waste a great deal of time, effort, and energy visiting them.

The golden rule is to develop the nearest area first, get it under control, and then expand the edges. Too often, the beginner chooses a garden far from the house, and neither harvests the plants efficiently, nor cares for them well enough. Any soil can be developed for a garden over time, so stay close to the home when placing the garden and orchard.

Zone 0 is the centre of activity (house, barn, or village if the design is on a large scale). It is laid out to conserve energy and to suit its occupants' needs.

Zone I is close to the house. It is the most controlled and intensively-used area and can contain the garden, workshops, greenhouse and propagation frames, small animals (rabbits, guinea pigs), fuels for the house (gas, wood), compost, mulch, clothesline, and grain drying area. There are no large animals on range, and perhaps only a few large trees (depending on shade requirements). Any frequently-visited or essential small tree can be placed in this zone, e.g. a reliably-bearing lemon tree.

Zone II is still intensively maintained, with dense plantings (larger shrubs, small fruit and mixed orchard, windbreaks). Structures include terraces, hedges, trellis, and ponds. There are a few large trees with a complex herb layer and understorey, especially small fruits. Plant and animal species that require care and observation are located in this zone, and water is fully reticulated (drip irrigation for trees). Poultry is let into selected areas (orchard, woodlot) to range, and an area for one milk cow can be fenced in from the next zone.

Zone III contains unpruned and unmulched

TABLE 1.1 SOME FACTORS WHICH CHANGE IN ZONE PLANNING AS DISTANCE INCREASES

Factor or Strategy	Zone I	Zone II	Zone III	Zone IV
Main design for:	House climate Domestic sufficiency	Small domestic stock and orchard	Main crop, forage,	Gathering, forage, forestry, pasture
Establishment of plants:	Complete sheet mulch	Spot mulch and tree guards	Soil condition and green mulch	Soil conditioning only
Pruning of trees:	Intensive cup or espalier, trellis	Pyramid and built trellis	Unpruned and natural trellis	Seedlings, thinned to selected varieties
Selection of trees and plants:	Selected dwarf or multi-graft	Grafted varieties	Selected seedlings for later grafts	Thinned to select varieties, or managed by browse
Water provision:	Rainwater tanks, well, bore, reticulation	Earth tank and fire control	Water storage in soils, dams	Dams, rivers, bores and wind pumps
Structures:	House/glasshouse, storage integration	Greenhouse and barns, poultry sheds	Feed store, Field shelter	Field shelter grown as hedgerow and woodlot

orchards, larger pastures or ranges for meat animals or rearing flocks, and main crop. Water is available only to some plants, although there are watering areas for animals. Animals are cows, sheep, and semi-managed birds. Plants include windbreaks, thickets, woodlots, and large trees (such as nut and oak) for animal forage.

Zone IV is semi-managed, semi-wild, used for gathering, hardy foods, unpruned trees, and wildlife and forest management. Timber is a managed product, and other yields (plant and wildlife) are possible.

Zone V is unmanaged or barely managed natural "wild" systems. Up to this point, we

design. In Zone V, we observe and learn; it is our essential place for meditation, where we are visitors, not managers.

Table 1.1 shows the factors which change in zone planning as distance increases.

Zones are a convenient, abstract way to deal with distances; however, in practice, zone edges will blur into each other, or landform and site access may mean that sometimes the least-used area (Zone V) is next to the most intensely-used area (Zone I); for example a steep forested hill directly behind the house).

We can in fact bring wedges of Zone V right

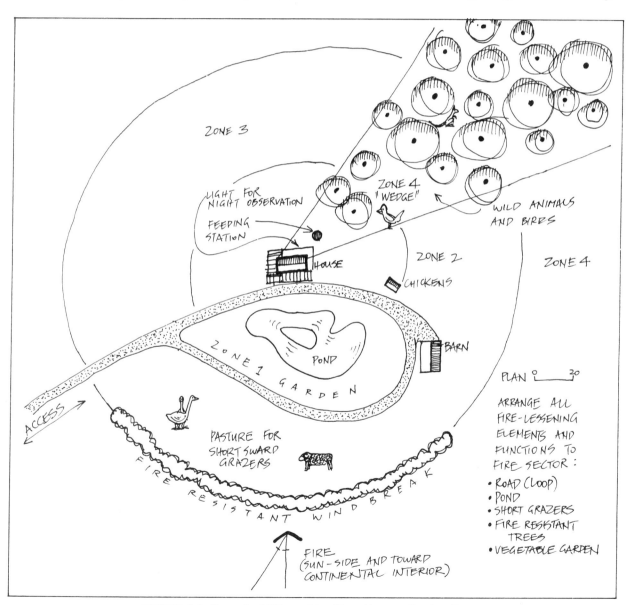

FIGURE 1.3 Fenced wildlife corridor (Zone V) extending into Zone 0.

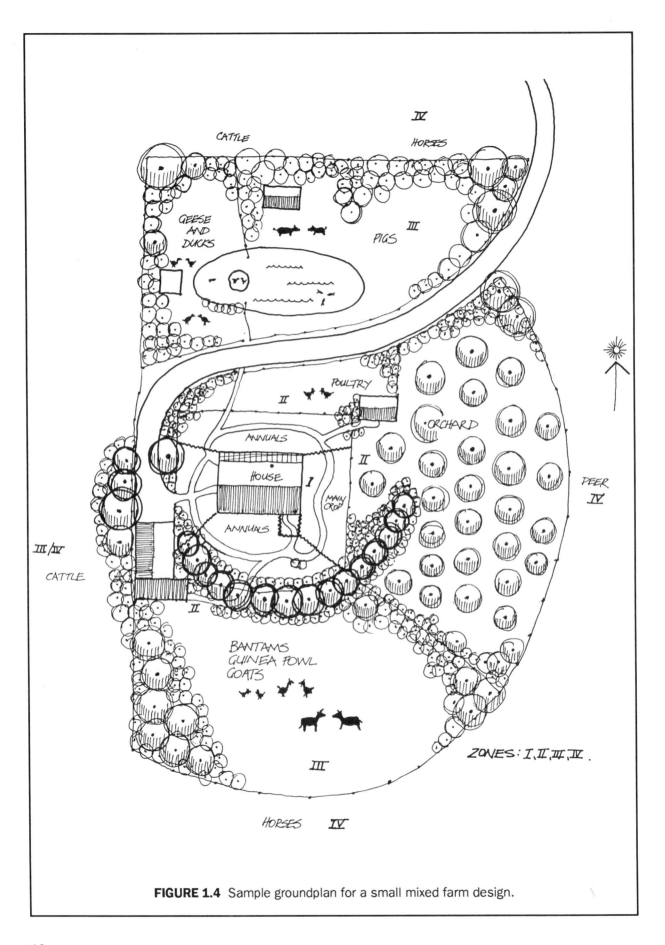

FIGURE 1.4 Sample groundplan for a small mixed farm design.

to our front door as a corridor for wildlife, birds, and nature. Or we can extend Zone I along a frequently-used path (a loop path which takes us from the house, to the barn, past the chicken shed, into the garden, near the woodpile, and back to the house). **Figures 1.3** and **1.4** show sample zone plans for a small farm.

Zonation patterns may change when we are working with two or more centres of activity, say between the house and a guest cottage, or the house and barn, or, on a larger scale, between the buildings in a village. In this case we must carefully work out linkages between these centres, consisting mostly of access, water and energy supply, sewage, and fencing connections. This is what David Holmgren calls "network analysis", which plans for more complex sites making connections between roads, pipes, windbreaks and so on to service more than one centre.

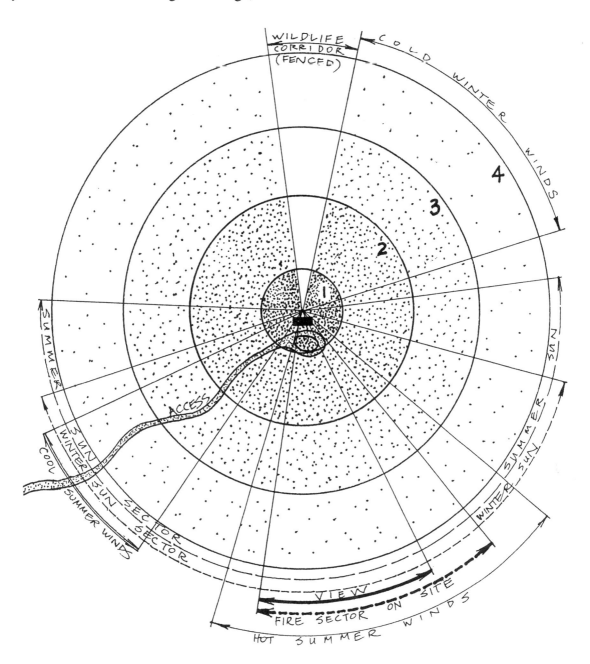

FIGURE 1.5 Understanding the direction from which sun, wind, fire and flood come helps in the placement of structures and vegetation.

Sectors deal with the wild energies, the elements of sun, light, wind, rain, wildfire, and water flow (including flood). These all come from *outside* our system and pass through it. For these, we arrange a **sector diagram** based on the real site, usually a wedge-shaped area that radiates from a centre of activity (commonly the house but it can be any other structure). **Figure 1.5.**

Some of the factors to sketch out on a ground plan are:
• fire danger sector
• cold or damaging winds
• hot, salty, or dusty winds
• screening of unwanted views
• winter and summer sun angles
• reflection from ponds
• flood-prone areas

We place appropriate plant species and structures in each sector (1) to block or screen out the incoming energy or distant view, (2) to channel it for special uses, or (3) to open out the sector to allow, for example, maximum sunlight. Thus, we place design components to *manage incoming energy* to our advantage.

For the fire sector, we choose components that do not burn, or that create firebreaks, such as ponds, stone walls, roads, clear areas, fire-

suppressing vegetation, or grazing animals to keep the vegetation short.

■ SLOPE

Finally, we look at the site *in profile*, noting relative elevations to decide on the placements of dams, water header tanks, or wells (above the house; waterfalls); to plan access roads, drains, flood or flow diversions; and to place wastewater or biogas units and so on. **Figures 1.6** and **1.7** illustrate some ideal relationships of structures and functions, given that there is a reasonable slope. Starting from the plateau or ridge:

• Dams placed above the house take overflow from high tanks, which rely on the roof catchment of hay storage sheds, workshops, or meeting halls, all of which need little water but have large roof areas for catchment. Diversion channels around high ridges leading to dams serve the same purpose.

• All covered tanks at high elevation are very useful, and these can in fact be built as the basement or foundation of the buildings, forming a heat/cold buffer in the sub-floor of workshops. Water from covered tanks is guaranteed free of biological pollution, and should be kept strictly for drinking at lower levels, the settlement area. Bulk domestic water (for showers,

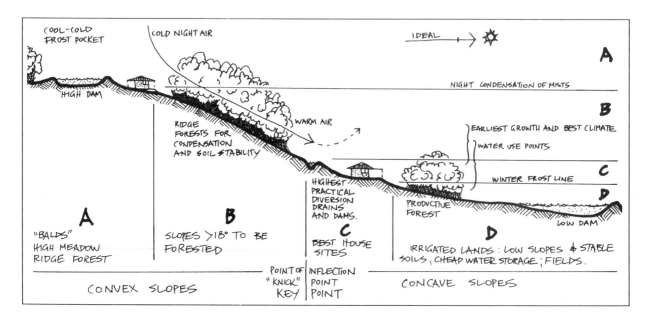

FIGURE 1.6 Slope analysis and site planning in relation to slope aspect largely decides the placement of access, water supply, forests, and cropland (for the humid landscape).

toilets, gardens) are supplied from the high dams.

• Above the house, particularly on rough, rocky and dry sites, there should be careful selection of dry-country plants needing "spot" watering only for establishment. These forests or orchards help with erosion control and water retention. On lower sites choose plants with higher water requirements.

• At the house, small tanks are needed for emergency water supply, and the house sited behind the lower dams or lakes for fire protection. Household greywater (wastewater from sinks and showers, not toilets), is absorbed by dense vegetation either in the garden or orchard.

• Downslope, water from the valley lake or large-volume storage at lower levels are pumped to the higher tanks or dams in emergencies such as fire or drought.

A factor often left unplanned is the high slope access, either as a track or road. Such access can provide water drainage or diversion to midslope dams, fire control on slopes, and harvest-time access to forest and to sheds or barns. Often enough, on small properties, the mulch from forests and manures from upslope barns can be easily moved downhill to establish a barn-to-house garden. Slatted floors in upslope shearing sheds, goatsheds, and stables allows easy access to manures.

To re-state the basic energy-conserving rules:

• Place each element (plant, animal, or structure) so that it serves at least two or more functions.

• Every important function (water collection, fire protection) is served in two or more ways.

• Elements are placed according to intensity of use (zones), control of external energies (sectors), and efficient energy flow (slope or convection).

Once this commonsense analysis is done, we know that every component is in a good place for three reasons (relative to *site resources*, *external energies*, and *slope or elevation*). To sum up, there should be no tree, plant, structure, or activity that is not placed according to these criteria. For instance, if we plant a pine tree, it goes in Zone IV (infrequent visits), *away* from the fire danger sector (it accumulates fuel and burns like a tar barrel), *towards* the cold wind sector (pines are hardy windbreaks), and it

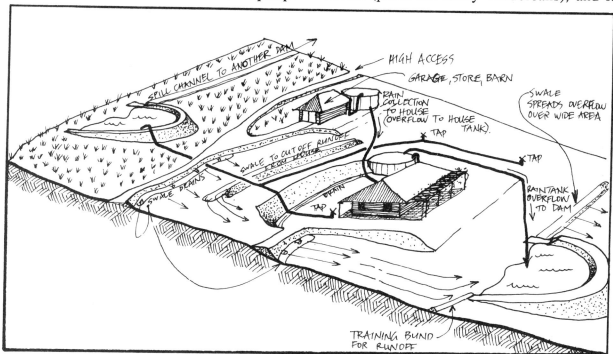

FIGURE 1.7 Idealised layout of water, buildings, and access (vegetation not drawn to better show water movements). Swales distribute water over a wide grassed slope to prevent gully erosion during wet seasons.

should bear edible nuts as forage.

If we want to place a small structure such as a poultry shed, it should *border* Zone I (for frequent visits), be *away* from the fire sector, *border* the annual garden (for easy manure collection), *back onto* the forage system, *attach* to a greenhouse in temperate climates, and form part of a windbreak system.

1.6
USING BIOLOGICAL RESOURCES

In a permaculture system, we use biological resources (plants and animals) wherever possible to save energy and to do the work of the farm. Plants and animals are used to provide fuel, fertiliser, tillage, insect control, weed control, nutrient recycling, habitat enhancement, soil aeration, fire control, erosion control, and so on.

Building up biological resources on site is a long-term investment which needs thought and management in the planning stages as it is *a key strategy* for recycling energy and developing sustainable systems. We use green manures and leguminous trees instead of nitrogen fertiliser; weeder geese and short herbs rather than lawnmowers; biological insect control rather than pesticides; and animals such as chickens or pigs instead of rotary hoes, weedicides, and artificial fertilisers.

However, careful and appropriate use of non-biological resources (fossil-fuel-based machinery, artificial fertilisers, technical equipment) in the beginning stages of a permaculture is OK *if* they are used to create long-term, sustainable biological systems and an enduring physical infrastructure.

For example, technological equipment such as photovoltaic cells, solar water heaters, and plastic pipes have used non-renewable resources in their manufacture, but we can use these effectively to produce our own energy on site. Similarly, we can hire earthmoving machinery to build roads, dams, swales and diversion drains; tractors to chisel-plough hard, unpro-

ductive ground or to disc-pit in drylands to trap silt and seeds for eventual plant growth; trucks to cart in manures and mulch from nearby sources so that we can get our own systems started.

In the same way, artificial fertilisers applied to worn-out soils will produce a green manure crop to start building up biological fertility. The problem comes when we are locked into an annual fertiliser or machinery treadmill instead of using these resources wisely to build up our own biological systems on site or in the community.

By all means carefully use what is available, use it for the best possible reasons, and develop alternatives as fast as possible.

Following are some examples of using plants and animals to increase yield and vigour, and to reduce the need for fertilisers and pesticides. Rather than relying on machines or brute force, we can instead *think* our way into managing and maintaining our properties.

Animal Tractors: Chickens and pigs are well-known for scratching and digging up the ground in search of worms, insects, and roots. Although animal tractor systems are described in Chapter 7, in brief chickens, pigs, or goats enclosed in a weedy or brambly area will destroy all vegetation, partly cultivate the earth, and manure the area. They are then rotated to another enclosure before they actually do damage through too much manure or soil disturbance.

Pest Control: Umbelliferous and composite plants such as dill, fennel, daisies, and marigolds placed around garden beds and in the orchard attract predator insects (insects which feed on or parasitise pests). Ponds in the garden attract insect-eating frogs. Suitable nest-boxes or thorny shrubs provide a habitat for insectivorous birds. Fungi and beneficial bacteria or nematodes have also been used to control insects, and many plants provide insect control or nematode control.

Fertilisers: All animals recycle nutrients by eating vegetation or other animals and excreting nitrogenous manures to fields, orchards, and gardens. Duck and pig manures in a large lake or pond increase the nutrients for many species of fish. Earthworms pump air into soils and provide humus and nutrients for plants, or are harvested for use as poultry and fish food. Garden and orchard wastes are recycled through worms, thus cleaning up potential pests and diseases.

Comfrey can be combined with manures and composted or fermented to a liquid mixture to provide essential nutrients for garden plants. Many vigorous and deep-rooted tree species probe the soil below the upper level and "mine" nutrients that are unavailable to more shallow-rooted plants. The leaves can then be used for mulch and to build up soil humus.

Legumes and leguminous trees (lucerne, beans, leucaena, acacias) which also provide nutrients to the soil by taking nitrogen from the air and processing it through root nodules, have a suitable bacteria (rhizobium) working in their root nodules. By adding the right rhizobium to potting soils, plant growth can be raised up to 80% over uninoculated individuals. (**Note**: not all legumes are nitrogen-fixing; notable exceptions are honey locust and carob). More than 150 non-leguminous plants, such as alder (*Alnus*), autumn olive (*Eleagnus*) and casuarinas, are also known to fix nitrogen.

Leguminous pasture, shrubs, and trees are interplanted amongst orchard and forest trees, and crop legumes such as broadbeans and field peas are planted in gardens and used as an understorey planting in orchards. If they are cut or pruned back before flowering, nitrogen from the root nodules is released into the soil, to be taken up by surrounding plants.

Many of these plants, especially the legumes, have other uses; the Siberian pea shrub (*Caragana*) and tagasaste (*Chaemocytisus palmensis*) for example, not only improve the soil, but are useful as a windbreak hedge, poultry food (seeds), and fodder for larger animals (leaves).

Other biological resources include bees (pollinating flowers and gathering nectar), spiny plants (fencing), allelopathic plants (plants suppressing weed growth), and dogs (livestock guard dogs, especially for sheep).

The key to using biological resources effectively is *management*. If not managed, such resources may become out of control and destructive, often ending up as pollutants. This can be seen as unfenced cattle eating forest saplings, escapee goats in the orchard, chickens polluting their pen; and untended leguminous trees shading out the garden.

Most management strategies are based on *timing*. For example, geese will weed the grasses from a garden containing strawberries, gooseberries, root crops such as onions and potatoes, tomatoes, etc. The key is to allow the geese into the garden *after* the plants are large enough (to prevent them from being damaged by geese feet) and *before* fruit ripens (geese will eat ripe strawberries and tomatoes).

Chickens, for all their advantages in manuring, and eating insects and weed seeds, should not be let into a mulched garden or orchard as they will scatter mulch while scratching around for insects. If the orchard is not mulched, but rather managed through an understorey of nitrogen-fixing legumes, chickens are let in to forage for fallen fruit, insects, and greens. Mulches in chicken pens can be covered with stones or with a metal mesh.

1.7 ENERGY CYCLING

In modern food-supply systems, full nutrition and a varied diet are provided by a world-wide transport, storage, and marketing network. This reticulation of food is, of course, more energy-expensive than local agricultural diversity and is only possible due to fossil fuel subsidy. Already, the costs of food reticulation are out of hand and are having their effects back at the farm. "Efficient" methods have been forced on the producer even if it is to the long-term detriment of the land or quality of the produce. Pesticides, large amounts of fertilisers, and un-

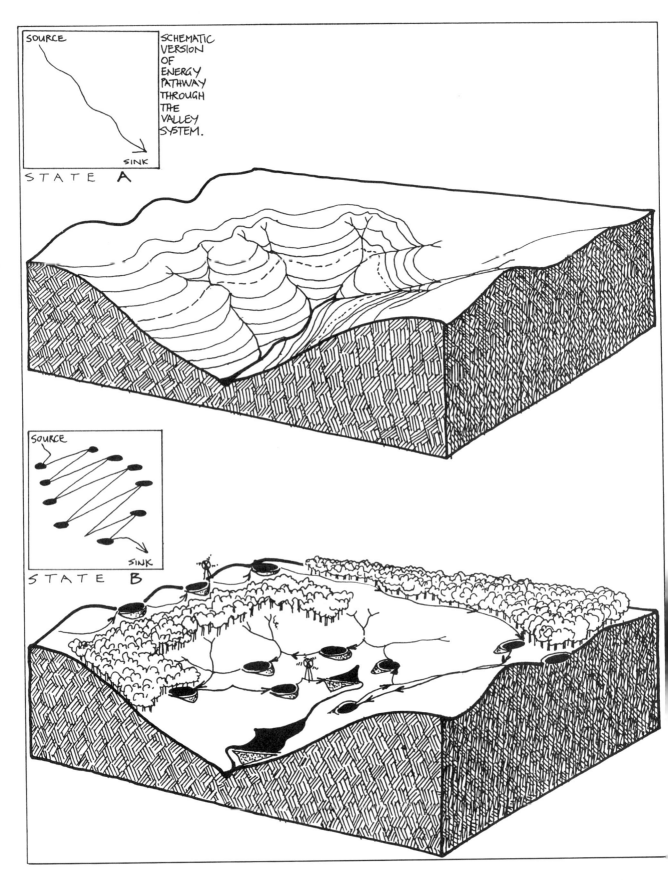

SOURCE

SCHEMATIC
VERSION
OF
ENERGY
PATHWAY
THROUGH
THE
VALLEY
SYSTEM.

SINK

S T A T E A

SOURCE

SINK

S T A T E B

FIGURE 1.8 The designer's work is to set up useful energy storages in a landscape or building (going from State A to State B). Such storages become resources for increased yields.

wise cropping sequences and cultivation techniques have become common-place in an effort to reduce costs and to raise yields in the hopeless race to remain economically viable.

A community supported by a diverse permaculture is independent of the distribution trade and assured of a varied diet, providing all nutritional requirements while not sacrificing quality or destroying the land that feeds it. The greatest savings on energy are in the elimination of costly transport, packaging, and marketing.

Permaculture systems seek to stop the flow of nutrient and energy off the site and instead turn them into *cycles*, so that, for instance, kitchen wastes are recycled to compost; animal manures are directed to biogas production or to the soil; household greywater flows to the garden; green manures are turned into the earth; leaves are raked up around trees as mulch. Or, on a regional scale, sewage is treated to produce fertiliser to be used on farmland in the district.

Good design uses incoming natural energies with those generated on-site to ensure a complete energy cycle.

The second law of thermodynamics states that energy is constantly degrading, or becoming less usable to the system. It is through constant cycling, however, that life on earth proliferates. The interaction between plants and animals actually *increases* available energy on the site. The goal of permaculture is not only to recycle and therefore increase energy, but also to *catch, store, and use* everything before it has degraded to its lowest energy use and so is lost to us forever. Our job is to use incoming energy (sun, water, wind, manures) at its highest possible use, then its next highest, and so on. We can create use points from "source to sink" before it runs off our property.

Water catchment and storage systems, for example, are constructed uphill for use in a complex pattern of ponds, smaller storages, energy generation, and so on, until the water is at last allowed to run off the property (**Figure 1.8**). If we ignore the hills and put a dam down in the valley, we have lost the advantage of

gravity and need energy to pump the water back up. It is not really the *amount* of rainfall that counts, it is the number of cycles we can set up to use that water to our best advantage. The more useful storages to which we can direct energy between energy coming into or generated on the site and energy leaving it, the better we are as designers.

1.8
SMALL-SCALE INTENSIVE SYSTEMS

Rather than large harvesters and transport trucks, a permaculture system is tuned for handtools (scythe, handmower, pruning shears, axe, wheelbarrow) on a small site, and modest fuel-users (tractor, mower, whipper-snipper, chainsaw) on larger sites.

Although permaculture may seem to be labour-intensive to start with, it is *not* a return to peasant systems of annual crops, endless drudgery, and total dependence on human labour. Rather it focuses on *designing* the farm (or garden, or town) to best advantage, using a certain amount of human labour (which can include friends and neighbours), a gradual build-up of productive perennial plants, mulching for weed control, the use of biological resources, alternative technologies that generate and save energy, and a moderate use of machinery, as appropriate.

Small-scale, intensive systems means that (1) much of the land can be used efficiently and thoroughly, and (2) the site is *under control*. On a small site, this is no problem; however, on larger sites it is easy to make the mistake of spreading out too quickly with extensive gardens, orchards, woodlot, and chicken-runs. This is a waste of time, energy, and water. If you want to know how to control your site, start at your doorstep. If you see a farm where the doorstep leads to weeds, then the weeds will go to the boundary; the land area is too big in terms of available time, labour, money, or interest.

If we cannot maintain or improve a system, we should leave it alone, thus minimising damage and preserving natural complexity. If we do not regulate our own numbers and appetites, and the area we occupy, nature will do so for us, by famine, erosion, poverty and disease. What we call political and economic systems stand or fall on our ability to conserve the natural environment. Closer regulation of available land, plus very cautious use of natural resources, is our only sustainable future strategy. Perhaps we should control only those areas we can establish, maintain and harvest by small technologies as a form of control on our appetites. This means that settlements should always include total food provision, or else we risk the fatal combination of sterile city and delinquent landscape, where city, forest and farm are all neglected and lack even the basic resources for self-sufficiency.

What we frequently observe in the western world is a delinquent landscape—the suburban plots under lawns and cosmetic flowers, and areas of urban blight around cities, more land cleared at the edge of the wilderness, and a desperate misuse of land in between. This system is not sustainable. At this moment, it seems clear that planning for highly-intensive, biologically-based food production at the doorstep is the only way out of future crises.

Contrast the large, cleared areas of Australia and North America with the small, intensively-farmed areas in the Philippines, where the total land around the house is usually only twelve square metres: out of this comes most of the food for the family. The house is often on stilts, with animals penned beneath. Garden surrounds the house. Scraps and trimmings are fed to the animals; manures are used on the garden. Trellis, holding passionfruit, gourds, beans, and other climbing vegetables, shelters the house from extreme heat and provides food for the family. Fast-growing trees (*Leucaena*) are coppiced for fuelwood.

So stay close to the house, and work towards developing small, intensive systems. We can plant ten critical trees, and look after them, whereas if we plant 100, we can lose up to 60% of them from lack of site preparation and care.

Ten trees and perhaps four square metres of garden, well protected, manured, and watered, will start the Zone I-II system.

The smaller nucleus plans are always in relation to a larger plan. They are the designs that surround the house, make up the orchard, or occur in the chicken run. The important thing to remember is to *fully develop the nucleus* before going on. The nucleus can be as simple as a large clump of pioneer trees, infrequently maintained, but established with good ground preparation and water provision if necessary. Or it can be a fully-planted, fenced, mulched, and watered garden, animal forage system, orchard, or pond margin. To save energy and water, and to prevent weed invasion, the developed system should be fully-occupied with plants, even if some will need to be thinned out later. Even if it appears to take more time and energy at first, it pays in the long-term through reduced plant death and easy system maintenance.

■ PLANT STACKING

In every ecosystem different plant species occur at varying heights above the ground, and root structures at different depths. Plants grow in response to available light, so that in a forest the mature trees form the uppermost (canopy) layer, with a lower tree stratum of smaller trees using some of the remaining light. The shrub layer, adapted to low light levels, grows beneath, and if there is any more available light, a herb layer forms as the lowest strata (**Figure 1.9**).

We can construct our own variation of the forest by establishing an intercrop of taller and shorter species, climbing plants, and herbs, placed according to their heights, shade tolerance, and water requirements. For example, on land with adequate fertility and a water source, we put our system in all at once, with climax species (long-lived orchard trees such as walnut or pecan); shorter-lived smaller fruit trees (plums, peaches); faster-growing leguminous pioneers (acacia, autumn olive, tagasaste) for mulch, shade, and nitrogen; short-lived perennials (comfrey, yarrow) to provide weed control

FIGURE 1.9 Plant stacking in a rich soil and water environment, sharing light and nutrients in canopy, mid-level, and herb strata.

and mulch; perennial shrubs (gooseberry, blueberry); and even annuals such as dill, beans, and pumpkin.

The spacing between plants depends mainly upon water availability and light requirements; dryland plantings require more spacing between them, while plants in hot, humid regions can be placed very close together. Cool climate design requires fairly open systems to allow light to lower layers and to overcome the lack of heat for ripening. Also, many temperate orchard trees and even plants in hot, humid environments need air movement between them to reduce the chance of fungal problems when unseasonal rains occur.

■ TIME STACKING

The British devised a system of farming in which pastures were broken up after the animals had been on them a few years. The proper rotation was every seven years. The pasture was plowed up and put into a high nutrient-demand crop, say lucerne, followed by a grain crop, followed by a root crop. One year it was left fallow to rest the soil. This was sustainable, but it took a long time to cycle. Masanobu Fukuoka, that master strategist, deals with *time stacking*. He does not have to fallow, because he never removes the main part of the crop from the soil. He stacks his legumes with his grains, with his ducks, and with his frogs. He sets his livestock in his crop at certain times instead of having a livestock site and a crop site. And he stacks different sorts of crops together. He goes one step further; he also stacks sequences into each other. He *starts* the next crop before the last crop is finished.

We can do the same thing by placing pioneers, young fruit trees, palms (or pole crop), shrubs, windbreak, ground cover, and even annual beds all together and at one time. Eventually annual crops will be shaded out by perennial shrubs and small trees, and in 20 years trees will dominate most of the area. Meanwhile, we will have harvested many years of produce and built up the soil through the addition of vegetable wastes and green manuring.

Instead of waiting for yields for 6-20 years from tree and nut crop, we get yields from 5-6 months on.

1.9
ACCELERATING SUCCESSION AND EVOLUTION

Natural ecosystems develop and change over time, giving rise to a succession of different plant and animal species. Abandoned pastures, for example, will be successively colonised by a weed and herb layer, pioneer plants, and eventually a climax species appropriate to soils, landform and climate. Each stage creates the right conditions for the next stage. Pioneer plants may fix nitrogen, loosen heavy soils, reduce salt in soils, stabilise steep slopes, absorb excess moisture, or provide shelter. They colonise new habitats, making it easier for other species to follow on by modifying the environment to a more favourable state.

Figure 1.10 shows the process of succession in a pasture system.

In conventional agriculture, vegetation is kept at the weed or herb level (e.g. vegetables, grains, legumes, pasture), using energy to keep it cut, weeded, tilled, fertilised and even burnt; that is, we are constantly setting the system back and incurring work and energy-costs when we stop natural succession from occuring.

Instead of fighting this process, we can *direct and accelerate* it to build our own climax species in a shorter time. We can do this by:

Using what is already growing, usually a "weed" layer, to build soil fertility. Soft weeds can be sheet-mulched with cardboard and old carpet, or slashed and used as mulch around other plants before seed heads develop. Woody perennial shrubs such as lantana and gorse make excellent soils when they break down after being slashed, and are eventually shaded out by forest trees. Roots may have to be dug out if we want a faster change, but for annual weeds, digging or turning under only produces more weeds as seeds sprout in response to light

FIGURE 1.10 EVOLUTION OF A DESIGNED SYSTEM

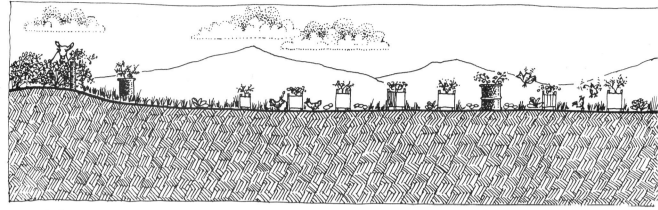

A. System establishment: an area is fenced and a mixture of species is planted and protected from grazers. Only geese, ducks, and some annual crops are harvested.

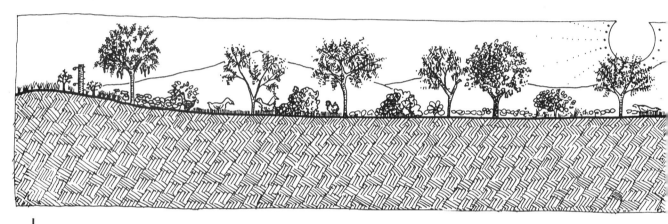

B. The system evolves to a semi-hardy stage. Chickens are introduced on an occasional basis.

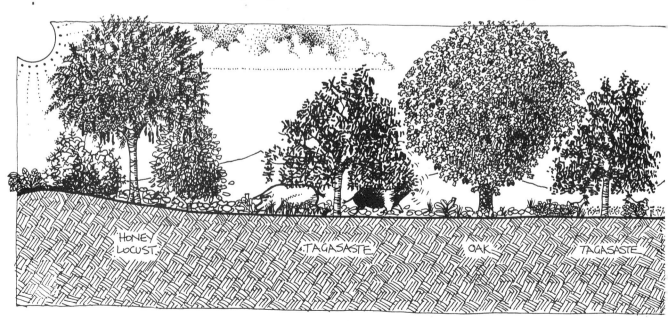

C. An evolved system provides forage, firewood, and animal products, and produces its own mulch and fertilisers. The mature system requires management rather than energy input, and has a variety of marketable yields.

and water.

Introducing plants that will easily survive in the particular environment and which will help to bring up soil fertility. Depending on the types of soils we are working with (which may be eroded, salted, swampy, worn out, acid, alkaline, clayey, or sandy), we can plant both annual and perennial types of a locally-adapted legume (for green manure and mulch), and shrubby useful perennials known to survive and thrive. We may need to wait to plant our "climax" crops until more favourable soils are established.

Raising organic levels artificially by using mulch, green manure crops, compost and other fertilisers to change the soil environment. This enables us to plant more quickly, or, if used in combination with the previous method, to plant a nucleus of climax tree crop in marginal ground if we are willing to put in the work of caring for those trees.

Substituting our own herb, pioneer, and climax species which are more useful to us than the existing natural or disturbed vegetation. Comfrey, for example, will come up through weed growth, helping to control the area if planted densely enough, and providing yields in the first year.

1.10
DIVERSITY

In his book *Plants, Man, and Life*, Edgar Andersen describes the garden/orchard plantings grouped around the houses in Central America. Close to the house and more or less surrounding it is a compact garden-orchard some 20 square metres in extent. No two of these are exactly alike. There are neat plantations more or less grouped together. There are various fruit trees (citrus, custard apple, sapote, mango and avocado), and a thicket of coffee bushes in the shade of the larger trees. There are tapioca (cassava) plants of one or two varieties, grown more or less in rows at the edge of the trees. Frequently there are patches of banana; corn

and beans are here and there in rows or patches. Climbing and scrambling over all are vines of various squashes and their relatives: the chayote (choko) grown for its squashes, as well as its big starchy root; and the luffa gourd, its skeleton used for dishrags and sponges. The cucurbits clamber over the eaves of the house and run along the ridgepole, climb high in the trees, or festoon the fence. Setting off the whole garden are flowers and various useful weeds (dahlias, rosemary, gladioli, climbing roses, asparagus fern, cannas and grain amaranth).

Andersen is contrasting the strict, ordered, linear, segmented thinking of Europeans with the productive, more natural polyculture of the dry tropics. The order he describes is a semi-natural order of plants, in their right relationship to each other (guilds), but not separated into various artificial groupings. It is no longer clear where orchard, field, house and garden have their boundaries, where annuals, and perennials belong, or indeed where cultivation gives way to naturally-evolved systems.

To the observer, this may seem like a very unordered and untidy system; however, we should not confuse order and tidiness. Tidiness separates species and creates work (and may also invite pests), whereas order integrates, reducing work and discouraging insect attack. European gardens, often extraordinarily tidy, result in functional disorder and low yield. Creativity is seldom tidy. Perhaps we could say that tidiness is something that happens when compulsive activity replaces thoughtful creativity.

Although the yield of a monocultural system will probably be greater for a particular crop than the yield of any one species in a permaculture system, the *sum of yields* in a mixed system will be larger. In the former, a hectare of vegetables will yield only vegetables throughout the year. In the latter, vegetables are a smaller part of the total yield of nuts, fruit, oil crop, timber, poultry, firewood, fish, seedcrop, and animal protein.

For self-reliance, this means that a family can satisfy all its nutritional needs with the available fruits, vegetables, proteins, and minerals. Economically, having more saleable products at different times of the year protects a family from market downturns and severe losses of one crop due to pests or bad weather. If the market for beef is down one year, for example, only firewood, nuts, fruit, seedcrop, and herbs are sold, keeping the cattle for better times. If frosts wipe out the fruit crop, other produce is available to eat or to sell.

Our aim should be to disperse yield over time, so that products are available during every season. This aim is achieved in a variety of ways:
- by selection of early, mid and late season varieties;
- by planting the same variety in early or late-ripening situations;
- by selection of long-yielding species;
- by a general increase in diversity or multi-use species in the system, so that leaf, fruit, seed and root are all product yields.
- by using self-storing species such as tubers, hard seeds, nuts or rhizomes which can be dug on demand;
- by techniques such as preserving, drying, pitting, freezing, and cool storage; and
- by regional trade within and between communities, or by purchasing land at different altitudes or latitudes.

Diversity is often related to stability in a permaculture. However, stability only occurs among *cooperative* species, or species that do each other no harm. It is not enough to simply place as many plants and animals as you can into a system, as they may compete with each other for light, nutrient, and water. Some plants, such as walnuts and eucalypts, inhibit the growth of others by releasing chemicals from their roots into the soil (allelopathy). Other plants provide overwintering habitat for pests and diseases harmful to nearby plants. Cows and horses grazed on the same pasture will eventually degrade it. Large trees compete with grain crops for light. Goats in the orchard or woodlot debark trees. Therefore, if we are to use all these elements in a system, we must be careful to place an intervening plant or structure between potentially harmful elements.

So the importance of diversity is not so much the number of elements in a system; rather it is the number of *functional connections* between these elements. It is not the number of things, but the number of ways in which things work. What we seek is a *guild* of elements (plants, animals or structures) that work harmoniously together.

■ GUILDS

Guilds are made up of a close association of species clustered around a central element (plant or animal). This assembly acts in relation to the element to assist its health, aid in management, or buffer adverse environmental effects.

We have long recognised companion planting in gardens, and crop mixes of various species in agriculture that do well together. Hence the concept of guilds which rely on composition and placement of species which benefit (or at least do not adversely affect) each other. Benefits can include:
- **Reducing root competition** from invasive grasses. Almost all cultivated fruit trees thrive in herbal ground covers, not grasses. Comfrey, for example, allows tree roots to feed at the surface and produces mulch and worm food when it dies down in winter, while spring bulbs (daffodils, *Allium* species) die down in the summer and do not compete with trees for water during summer-dry periods.
- **Providing physical shelter** from frost, sunburn, or the drying effects of wind. Examples are hedges and borders of hardy trees and shrubs which deflect strong winds, and scattered trees which provide partial shade for crops such as coffee and cocoa.
- **Providing nutrients** in the form of leguminous annuals, shrubs, or trees.
- **Assisting in pest control** by providing chemical deterrents (*Tagetes* marigolds fumigate the soil of certain types of nematodes); hosting insect predators (umbelliferous plants

such as dill, carrot, and fennel); and using animal foragers such as chickens to clean up fallen fruit.

It is this last item which interests me in regard to pests in the garden, orchard, and cropland. Plants can be defined as interacting positively or negatively. Of great importance in crop mixtures are pest interactions and functions of the plant species involved:

• **Insectary plant**: the plant acts as a host (a food plant) for predatory insects which prey on the crop pests.

• **Sacrificial plant**: pests attack this plant preferentially, which nevertheless does not prevent it setting seed. Other plants nearby escape severe predation.

• **All-season host** : pests overwinter or live in this type of plant, enabling them to build up larger populations (e.g. pests of citrus are hosted off-season by oleanders).

• **Predator or pollinator attractor plant**: the crop or hedgerow species provides flowers to feed the adult predators (e.g. buckwheat in or near a strawberry crop).

• **Trap crops**: some crops can attract and kill pests, or the pests can be caught or destroyed on those crops.

These important functions are served by trees, shrubs, flowers, and vines, so that any farmer who carefully selects hedgerow species to be in one or more of the above categories has substantial pest control capabilities.

If we have a system with diverse plant and animal species, habitats, and microclimate, the chance of a bad pest situation arising is reduced. Plants scattered amongst others make it difficult for pests to go quickly from one food plant to another. However, once pests do breed on any one tree, insect predators perceive this as a concentrated food source and will congregate to take advantage of it. In the monocultural situation the food for pests is concentrated; in a polyculture, the pest itself is a concentration of food for predators.

1.11
EDGE EFFECTS

An edge is an interface between two mediums: it is the surface between the water and the air; the zone around a soil particle to which water bonds; the shoreline between land and water; the area between forest and grassland. It is the scrub, which we can differentiate from grassland. It is the area between the frost and non-frost level on a hillside. It is the border of the desert. Wherever species, climate, soils, slope, or any natural conditions or artificial boundaries meet, we have edges.

Edges are places of varied ecology. Productivity increases at the boundary between two ecologies (land/water; forest/grassland; estuary/ocean; crop/orchard) because the resources from both systems can be used. In addition, the edge often has species unique to itself. In nature, reef ecologies (the edge between coral and ocean) are some of the most highly productive systems in the world, as are mangrove ecologies (land/sea interface).

There is hardly a sustainable traditional human settlement that is not sited on those critical junctions of two natural economies, here the area between foothill and forest and plains, elsewhere on the edge of plain and marsh, land and estuary, or some combination of all of these. A landscape with a complex edge is interesting and beautiful; it can be considered the basis of the art of landscape design. And most certainly, increased edge makes for a more productive landscape.

Planners who place a housing settlement on a plain may have the "advantage" of plain planning, but abandon the inhabitants to failure if transport fuels dry up, when they will have to depend on a limited natural environment for their varied needs. Successful and permanent settlements have always been able to draw from the resources of at least two environments. Similarly, any settlement which fails to *preserve* natural benefits, and, for example, clears all forests and poisons estuaries, rivers or soils, is bent on eventual extinction.

26

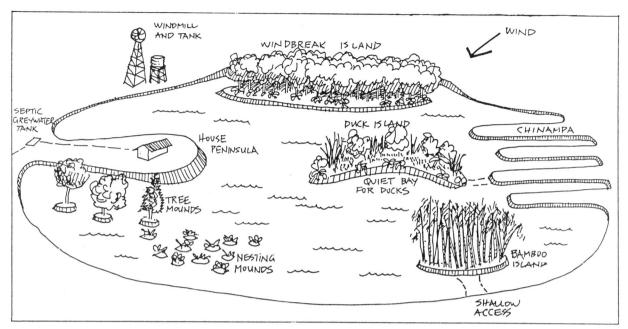

FIGURE 1.11 Useful earthbanks and islands in and around water provide many edges (niches) for plants, animals, and people.

We can either place our homes and settlements to take advantage of the resources of two or more ecosystems, or we can increase the complexity of our properties by designing and constructing our own varied ecosystems. If we haven't settled near water, we can make dams and ponds; if we are on flat land, we can use machinery to shape soil mounds or bunds up around us; if we have no forest, we can grow a woodland, no matter how small. Even within the larger property, we can think in terms of "edges" for smaller elements. For example, a pond can either be of one shape and depth (and host a simple ecology), or we can build it with varying depths, shapes and islands. Then we can plant rushes at the edge of the pond, waterlilies and water chestnut in the shallows, and have carp feeding on fringing vegetation at the top of the pond, catfish roaming the pond bottom, and birdlife on a sheltered island (**Figure 1.11**).

The edge (boundary) acts as a net or sieve: energies or materials accumulate at edges, e.g. soil and debris are blown by wind against a fence; seashells form a line at the tide-marks on a beach; leaves accumulate at kerbsides in a city. By noting how edges trap materials in nature, we can design to take advantage of the natural drift of materials or energies in our system. People who build roads in snow country recognise the value of constructing special lattice fences to intercept snow so that it does not settle on the road. In desert country, where mulch is scarce, we can construct "mulch traps" in stream beds simply by adding a large log or fence at an angle to the stream. During floods the debris (silt and vegetation) carried by the water is deposited just before or after these logs or fences.

Edges define areas, and break them up into manageable sections. Edges can be defined along fencelines, access roads, pond shores, the area between the house and the driveway, the path around the garden, terraces, and in fact any area that can be defined by a structure (fence, trellis, house, or chicken run), access (walkway, path or road) or a line of vegetation (windbreak or barrier hedge). So edge is also important in permaculture from the standpoint of implementing and maintaining a section of the designed system.

Only by defining the edges around an area can we begin to control it. If we do not control the edge around our garden by planting barrier

27

plants and weed suppressors, elements from outside the garden (animals, weeds) will invade it. In addition, we walk at the edge and we pause there; our energies are devoted to species we have access to rather than to those which may be in the middle of a large expanse of unbounded territory.

Now we come to the concept of edge in a different way: from its geometry, or *pattern*. Think about the configuration of our brains, our intestines. There are yards of material packed into a small space, and a lot of edge or function possible. Perhaps we too can increase the yield of our system by manipulating the *shape of the edge*. A curved edge can be more useful than a straight edge, particularly if the curve also spirals up. A wavy (crenellated) edge is more useful still, allowing access to more area. Mounds or earthbanks also show a lot of edge; more plants can be placed on a spiral ramp around a mound, especially in a small garden space. So let's see what can be done when we play around with edge configurations.

Spiral: When we make our garden beds, we usually get the string and rake everything out and make it level. If the garden wasn't level to start with, we soon level it. But what if our gardens went up into the air, or even down into the ground? The shape of a type of seashell that spirals up is a very efficient way of stacking a lot of digestion into a little space. A herb spiral is just this (**Figure 5.1**). The base is 1.6 metres across, with a planting ramp spiralling up the middle. Herbs are planted into the spiral according to their needs, with sun-loving herbs facing the sun, and shade-loving ones on the other side. With just one move, we condensed space, created a variety of micro-climates, increased edge for greater yield, and relieved the monotony of a flattened landscape.

Lobular or Crenellated: I used to live by the sea and my trees were always getting blasted by the wind. However, over the road I had a large clump of spiny boxthorn (*Lycium ferrocissimum*) and one day I got the brushhook and cut out a complex series of bays (**Figure 1.12**), leaving the perimeter intact for protection against

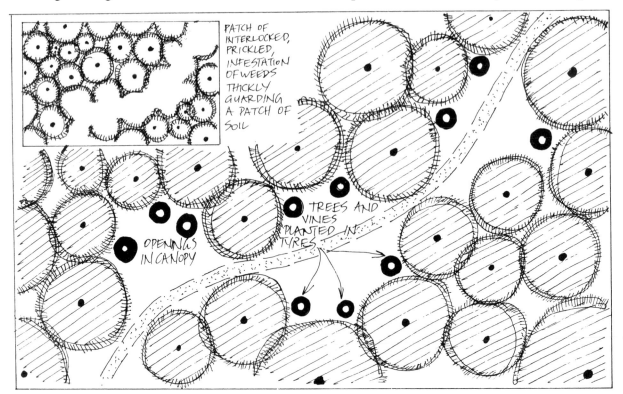

FIGURE 1.12 Lobular pattern cut through a large bramble area (e.g. boxthorn, blackberry, or gorse) will protect trees from browsing animals and wind-blast, especially on sea coasts. Old tyres around trees protect plants from ground browsers such as rabbits. One dripline services all trees.

wind and cows. Now I had a variety of micro-climates: warm spaces, areas with cold winds, shady spaces, and dry and wet areas. And I had a lot of edge in which to plant: so I put in my fruit trees surrounded by a smaller herb layer of marigolds and comfrey. One dripline watered the area, and for mulch around the trees, I slashed a bit more of the growing boxthorn.

A crenellated form (large or small lobes) gives far more edge than a straight line (**Figure 1.13**), and hence more yield. The round pond on the left has exactly the same area as that on the right, but yield has doubled due to more water/land edge.

Chinampa: The chinampa system of Mexico and Thailand consists almost entirely of edge (**Figure 1.14**). These ditch-and-bank configurations are highly-productive systems; plants growing on the bank have access to water, and fish in the ditch make use of the fringing vegetation. Muck from the bottom of the ditch is brought up in buckets and used to keep the garden beds on the bank fertile.

Edge Cropping: Edge cropping has been used extensively in many parts of the world where two crops (e.g. wheat and lucerne, tree crop and row crop) are planted in strips. We can develop more complex systems (**Figure 1.15**) by planting strips of tree crop, comfrey (a permanent mulch and nutrient plant), legumes (either for harvest or for use as a green manure), sunflowers (for human or animal food), and vegetables. The vegetable residues (sunflower and corn stalks) are used as mulch and

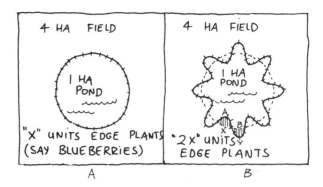

FIGURE 1.13 Without altering the area of the field or the pond, we can double the plants on the pond-edge by changing the shape of the edge to increase the earth/water interface.

nutrient for the trees. Harvesting and maintenance is greatly assisted by contoured tracks or cropping in strips.

In tropical areas, a system of *avenue cropping* employs a legume tree (leucaena, sesbania, *Cajanus* spp., *Acacia* spp., gliricidia) which is planted in strips with vegetable crop (maize, pineapple, sweet potato). The legume tree, coppiced or used as shade annually, provides nitrogen and mulch to the crop. It also produces firewood (**Figure 5.10**).

Edge patterns can be zigzag (zigzag fences stand up to wind better than straight fences); lobular (keyhole beds create different micro-climates); elevated (mounds and banks provide wind protection, greater growing surface, and good drainage); pitted or "waffle iron" (for garden beds in dry climates, and to trap mulch and debris blowing across the landscape); gently curved (paths cut on the contour along hill-

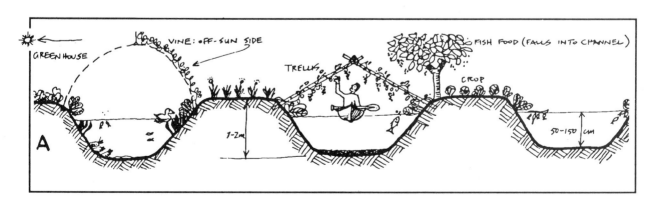

FIGURE 1.14 Ditch and bank (chinampa) systems are highly productive

sides allow access for planting, mulching and watering); and sharply curved (suntrap design to enhance heat and protect from cold winds). **Figure 1.16** shows some types of edge patterns.

We need to select appropriate edge patterns for climate, landscape, size, and situation, as different kinds of systems and plant species need different approaches. Small-scale systems allow greater pattern complexity; large-scale systems must be simplified to minimise work.

1.12
ATTITUDINAL PRINCIPLES

The foregoing ideas are environmental and permacultural principles. They deal with the site, or the environment, or the actual design. The following are people-oriented principles, and deal with principles of *attitude*.

■ EVERYTHING WORKS BOTH WAYS

Every resource is either an advantage or a disadvantage, depending on the use made of it. A persistent wind coming off the sea is a disadvantage for growing crops, but we can turn it into an advantage by building a wind-generator and placing our garden within shelterbelts or in a greenhouse.

Disadvantages can be viewed as "problems" and we can take an energy-expensive approach to "get rid of the problem", or we can think of everything as being a positive resource: it is up to us to work out just *how* we can make use of it. "Problems" can be intractable weeds (e.g. lantana in the tropics), huge boulders lying on the perfect house site, and animals eating garden and orchard produce. How can we turn these into useful components of our system? Lantana is an excellent soil builder; it can be shaded out with a vigorous vine such as choko (chayote), or slashed and used as rough mulch around pioneer trees (which will eventually shade out the lantana if planted densely). Boulders on the perfect house site can be incorporated into the house itself, for beauty and as a heat storage system. Animals can be trapped and eaten; blackbird pie was a standard favourite in England with good reason; possum skins are warm; and venison is undoubtedly a better protein than beef.

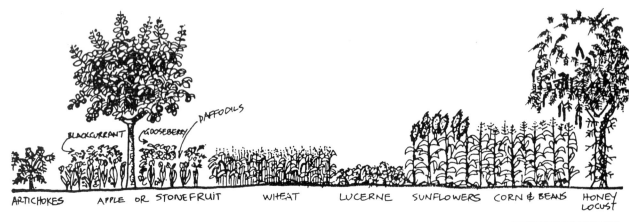

ARTICHOKES   APPLE OR STONEFRUIT   WHEAT   LUCERNE   SUNFLOWERS   CORN & BEANS   HONEY LOCUST

FIGURE 1.15 Edge cropping in orchard and field crop. Note that field (A) and field (B) at right are of equal area, with the same inter-row and in-line spacing. However, in field (B) we can fit 45 plants as opposed to field (A), with only 36 plants.

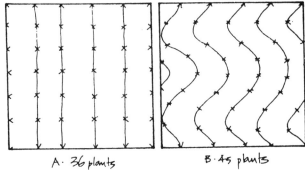

A. 36 plants   B. 45 plants

■ PERMACULTURE IS INFORMATION AND IMAGINATION-INTENSIVE

Permaculture is not energy- or capital-intensive, rather it is information-intensive. It is the quality of thought and the information we use that determines yield, not the size or quality of the site. We are using not only our physical resources, but our ability to access information and to process it.

Information is the most portable and flexible investment we can make in our lives; it represents the knowledge, experience, ideas, and experimentation of thousands of people before us. If we take the time to read, observe, discuss, and contemplate, we begin to think in terms of multidisciplines, and to design systems which save energy and give us yields.

The yield or living that can be made from a particular site, for example, is not limited by

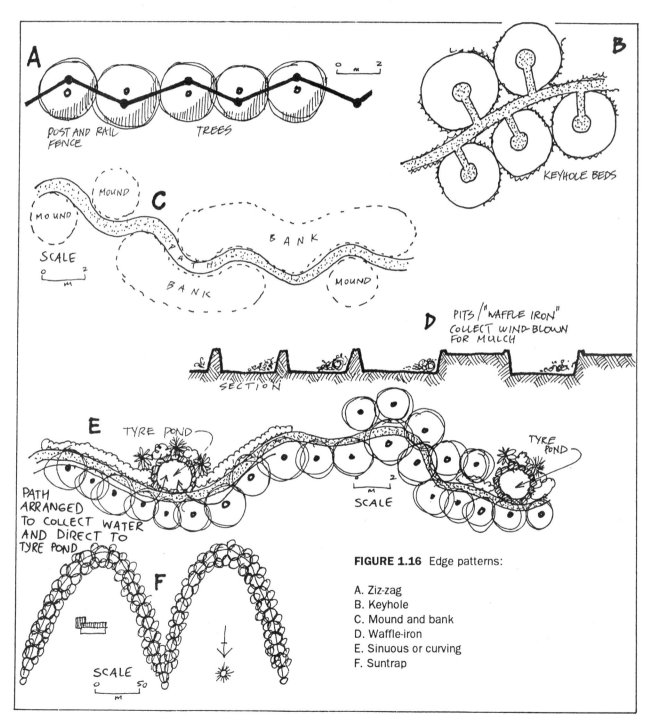

FIGURE 1.16 Edge patterns:

A. Ziz-zag
B. Keyhole
C. Mound and bank
D. Waffle-iron
E. Sinuous or curving
F. Suntrap

size, but rather by how effectively we can use a particular niche. It is the number of niches in a system that will allow a greater number of species to fit into our design; our job is in working out how we can create them. For example, the number of pairs of pigeons breeding in a cliff depends on the number of ledges. If we want pigeons on our property (for their manures, or to eat), we can provide more ledges in the form of pigeon houses around the garden. We see how things work in nature, and take our ideas from there.

Even if we have an energy-efficient property (where the waste products of one element are used for the needs of another element), fully-planted, and under control, there is always some better way in which it can work, always another niche to fill. The only limit on the number of uses of a resource possible within a system, is in the limit of the information and the imagination of the designer.

1.13
REFERENCES AND FURTHER READING

Anderson, Edgar, *Plants, Man and Life*, University of California Press, Berkeley, 1952.

Kern, Ken, and Barbara Kern, *The Owner-Built Homestead*, Charles Scribners's Sons, 1977

Odum, Eugene, *Fundamentals of Ecology*, W.B. Sauders, Toronto, 1971.

Phillbrick, N., and R.B. Gregg, *Companion Plants*, Robinson and Watkins, London, 1967.

Quinney, John, *Designing Sustainable Small Farms*, Mother Earth News (July/August 1984)

Whitby, Coralie, *Eco-Gardening: The Six Priorities*, Rigby Pub. Ltd., 1981.

Broadscale Site Design

2.1 INTRODUCTION

This chapter focuses on site design in a broad sense: analysing resources; working with the site limitations of landform, microclimate, soils, and water; and siting the house, access, and fencing for maximum benefit and to avoid catastrophes such as fire and flood.

Planning the design is the single most important thing we can do before putting anything in place. The overall plan, if done thoroughly, will save time, money, and needless work.

There are several ways to start the design process, depending on your nature and needs. You can start out by defining your goals, as precisely as possible, and then look at the site with these goals in mind. Or you can take the site with all its characteristics (both good and bad), and let goals suggest themselves. Of the two questions—"What can I make this land do?" or—"What does this land have to give me?"— the first may lead to exploitation of land without regard to long-term consequences, while the second to a sustained ecology guided by our intelligent control.

Defining goals and identifying site potentials and limitations go hand-in-hand. It is always easier to see the site with goals in mind, even if those goals later turn out to be unrealistic. In fact, goals may need to be re-defined in view of site limitations. Design is a continuous process, guided in its evolution by information and skills derived from earlier experience and observations. All designs that involve life forms undergo a long-term process of change; even the "climax" state of a forest is an imagined concept.

2.2 IDENTIFYING RESOURCES

Observation and research are used to identify the resources and limitations of a particular site. We get maps of the property, and consult records of wind, rainfall, flood, fire, and species lists of the area. We ask local people about pests, problems and techniques they use. This information gives us a broad picture of the area. They set the scene. However, they tell us nothing about the site itself. Only by walking the site and observing it in every season can we discover its limitations and its resources. We can change many of these over time by good design, appropriate plant and animal species, water storage, windbreaks, and so on.

■ MAPS

A good feature map is a great help in site design; it reveals waterways, vegetation, soils, geology, and access (all essential or useful information). We can make a map or buy one, and combine several specialist maps or aerial photos to picture the site. If the maps show good contour lines, these can help us to design water systems and to place components that need specific aspect, slope, or altitudinal advantage.

Things to map are the natural features that include landform (size, shape, geological features, slope, and aspect), existing vegetation, streams, and soils; and built environment ("improvements") such as fences, roads, buildings, dams and earthworks, power and water connections, etc. If we walk around the site and colour in all these factors on the map, the site almost begins to design itself. Planted trees, pastures or windbreaks may be considered either as part of the natural or the built environment depending on whether they are clearly recent improvements or long-established evolved parts of the landscape.

Maps are useful only when they are used in combination with observation. Never try to design a site by just looking at a map, even if it is thoroughly detailed with contour lines, vegetation, erosion gullies, and so on marked in. Maps are never representative of the complex reality of nature. Obtain good maps if you can, but pay more attention to the ground, to the behaviour of organisms, pioneers, water, wind, and to seasonal changes. Remember, "The map is not the territory." (Korzybski, *General Semantics*).

■ OBSERVATION

As we walk about a site and talk to people, we can note our observations. At this stage, we try to store the information we gain in some accurate way, carry a notebook, or a camera and tape-recorder, and make small sketches. The notes we end up with can later be used to devise design strategies.

We do not just see and hear, smell and taste, but we sense heat and cold, pressure, stress from efforts of hill-climbing or prickly plants, and find compatible or incompatible sites in the landscape. We note good views, outlooks, soil colours and textures. In fact, we use (consciously) all our many senses and become aware of our bodies and responses.

Beyond this, we can sit for a time and notice *patterns* and *processes*: how some trees prefer to grow in rocks, some in valleys, others in grasslands or clumps. We see how water flows on the site, where fires have left scars, winds have bent branches or deformed the shape of trees, how the sun and shadows move, and where we find signs of animals resting, moving, or feeding. The site is full of information on every natural subject, and we must learn to read it well.

Reading the landscape is a matter of looking for *landscape indicators*. Vegetation in particular provides information about soil fertility, availability of moisture, and microclimate. Rushes, for example, indicate boggy soils or seepages; dandelions and blueberries acid soils; and dock compacted and clayey soils. Large trees growing in dry regions indicate a source of deep water. An abundance of thorny or unpalatable weed species (thistle, oxalis, sodom apple) indicates overgrazing or mismanagement; erosion gullies and compacted pathways will confirm this. A plant flowering and fruiting earlier than others of the same species indicates a favourable microclimate, and trees growing with most of their branches on one side indicates the direction of strong prevailing winds.

These examples are specific to different climates and even to different landscapes. Locally-developed rules of thumb come from knowledge of the local region.

Fire frequency and direction can also be seen by noting changes in vegetation. Fire produces dry, scrabbly, summer-deciduous, thick-seeded species; lack of fire develops broadleaf, evergreen or winter-deciduous, small-seeded plants and a deep litter fall. Often, trees and other plants will indicate frost-lines on slope properties by a change in vegetation type.

As we observe, we can note potential "problems", such as noxious vegetation, erosion gullies, boggy ground, rocky areas, or compacted, leached soils. These are areas of special consideration, and might be selected for special yields, or left untouched for wildlife areas. Some problems, with a little thought, are turned to advantage. Boggy ground is an indicator of the natural drainage patterns of the area, and reflects impermeable subsoils; these can be made into a wetland area, or dug out to provide open water. Sometimes below swamps and marshes there is an accumulation of peat, or high-value potter's clay. If ponds are made in the marsh, some peat can be harvested for potting soil or to improve sandy areas.

There are many resources to look for. Are there streams or water sources at head (for water supply and possible energy supply)? Are there forests containing valuable timber, or even dead logs useful for wildlife or for firewood? Is there a good wind site for wind power?

There are many categories of resources: earth resources; biological resources (plant, animal, and insect life); the energy resources of wind, water, wood, oil crops and gas; and the social resources. The social resources include the potential of the site for teaching and seminars, or recreational activities, which depends mainly upon location, facilities available or that can be constructed, and local planning laws.

By observing the landscape we draw inspiration from the survival strategies followed by natural systems, and imitate them using species of more direct use to us. We observe, for example, that large trees grow on the shade side of deep dryland canyons; this is where we can place our own trees for assured success. Or we see that pioneer plants are establishing at fencelines and posts from bird droppings; we can set dozens of perch posts throughout the landscape to encourage such plants, or we set out perches near small fruit trees to provide phosphate for our trees.

■ **OFF-SITE RESOURCES**

We can find out about opportunities in the local area. Sawmills, dumps, markets, horse stables, restaurants, and chicken farms are all potential resources; waste products can be used to improve the site while our own resources are being developed.

One of the most overlooked factors is *access*

to non-site resources, e.g. shops, schools, markets, and other services. Estate agents recognise the value of locating closer to towns, with land prices higher the closer they are to essential services. While permaculture places more emphasis on site resources, external resources are often critical not only in establishing a system, but in the time and money it takes to get to town (for work or school). Parents far from a main highway will often need to travel twice a day to deliver and pick up school children.

It is also important to take your own resources into account. Are your skills and financial assets equal to the design you would like to implement? Can your skills and products be used in the local area? Is there a market for fine herbs, nursery stock, free-range poultry, organic fruit and vegetables, seeds, water lilies, freshwater fish, or whatever your permaculture system can provide? Can you use local revolving funds to assist changes, given a realistic business plan?

2.3
LANDFORM (TOPOGRAPHY)

Topography or landform is an unchangeable feature of a site, and although minor earthworks can alter some of the nature of the site, large earthworks are expensive and usually unnecessary.

Topography has an effect on the microclimate, water drainage patterns, soil depth and character, access, and view of a site. To understand its influence on the land, the topographical features that must be noted and mapped are:
   • Sun-facing and shade-facing slopes;
   • Cliffs or rocky outcrops;
   • Drainage lines (watercourses);
   • Rough terrain;
   • Good and bad views;
   • Hill heights, gradients, and access;
   • Boggy areas, areas susceptible to erosion, and so on.

Obviously a small site will be easy to map,

while a large acreage will take some days or weeks.

A variable site with many of the above features is most useful, especially in regard to slope. Slopes are noted as to aspect (whether it faces north, south, east or west), and gradient (gentle, medium, or steep), the latter being a good indication of potential erosion problems, especially if trees have been cleared off a steep hill. The effect of slope on microclimate is discussed in the following section.

It is important to note that permaculture can be developed on any type of country: rocky hills, swamps, alpine regions, alluvial river flats, or deserts. It is not necessary to try to change a stable landscape to achieve particular conditions, as every landscape and natural ecosystem will dictate the general nature of the permaculture possible; this is necessary if the system is to have long-term viability.

2.4
CLIMATE AND MICROCLIMATE

Climate is the basic limiting factor for plant and animal diversity in an area. Although any site planning must consider the overall climate of the region (humid-hot, dry-hot, arctic, temperate, etc.), we must take particular note of the different microclimates due to topography, soils, vegetation and other factors. Two properties, located only a few miles apart, can vary in rainfall, wind speed, temperature, and relative humidity, so it is vital to analyse the site climate in detail rather than to rely on the broad climatic statistics for the district. This important basic step can mean the difference between living in pleasant surroundings or in miserable conditions on a property that will probably change hands every few years.

If we study the microclimates on our site, we will be able to:

• Place structures, plants, and animals in the most favourable sites, (e.g. the house facing the sun in temperate climates, or the house on the shade side of a hill in hot climates);

• Focus beneficial energies and scatter hostile energies coming into the site (e.g. plant wind barriers near the house and crop, or conversely, plant trees in such a way as to funnel breezes towards a house);

• Extend favourable microclimates.

The next sections discuss factors which can most affect microclimate on a site, and so should be considered with the house site and growing areas in mind.

■ TOPOGRAPHY

Topography refers to the landscape features of a site, usually to what degree it is hilly or flat. Flat areas will have very little difference in topography (meaning little or no differences in microclimate), whereas hilly areas show great variation in microclimate.

Aspect

Aspect refers to how slopes are oriented in relation to the sun, and affects site conditions due to the amount of direct sun they receive. Slopes facing "sunwards" (north in the southern hemisphere and south in the northern hemisphere) receive the most sun; if they are also facing east, the maximum temperature is reached in the morning, while if they are facing west, the maximum temperature is reached in the afternoon. A slope facing "the shade side" (north in the northern hemisphere and south in the southern hemisphere) will receive very little direct solar radiation.

The influence of aspect on plants in natural plant communities can be seen where sun-facing slopes are covered by dry sclerophyll forest, while on the cooler, wetter shade-facing slopes, they may be occupied by wet sclerophyll forest (**Figure 2.1c**). The use of aspect in permaculture usually means taking advantage of the sun-facing slopes since these are useful for ripening fruits, siting the house for the most warmth during winter, and planting vegetation that is "marginal" for that particular climate, such as a tropical tree in a subtropical region.

Conversely, plants or structures that need shade or additional coolness are placed on shade-

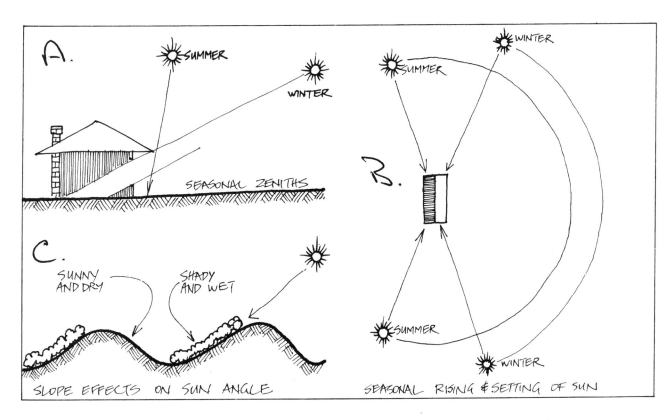

FIGURE 2.1 Sun direction and its seasonal height affect house design and plant communities.

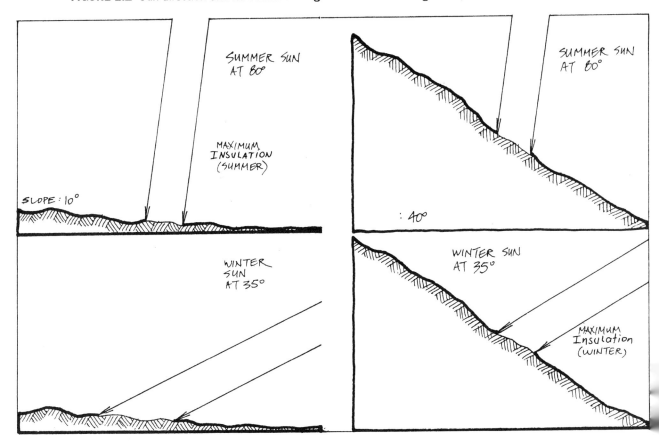

FIGURE 2.2 How slope affects the amount of direct solar radiation received at different seasons.

side slopes, e.g. a cool cellar for storing wines, or cool-climate berries in subtropical climates.

For energy-efficient house design in particular, but also for placing gardens and orchards, it is essential to note the seasonal variation of the sun's path, particularly its absolute height in the sky between summer and winter (**Figure 2.1a**) and the distance it travels on its path from east to west (**Figure 2.1b**).

Aspect is not as important a factor in cloudy climates, or when the sun is shaded by even larger topographic features, such as a mountain or ridge opposite the site.

The effect of *aspect* plus the actual steepness of the slope is quite marked. As can be seen in **Figure 2.2**, a gentle slope is warmer in the summer as it receives sunlight coming in at a favourable angle. However, the best slope for winter is a steep one, as it receives the sun at a better angle than a gentle one.

Cold Air Drainage

The steepness of the slope affects both water runoff and soil stability, but in terms of microclimatic planning, it mostly influences cold air drainage. Cold air is heavier than warm air, and tends to flow from convex hills into concave valleys. It will pool in the valleys, increasing the likelihood of frost. Hilltops are also prone to frost as pools of cold air remain on flatter ridge tops and plateaus. The most frost-free sites are usually on the upper or mid slopes of valleys above 20 metres. Because they are warmer, night and day, than either valley floor or ridge, such areas are known as the *thermal belt* (**Figure 2.3**). This area has long been used for siting villages, houses, and favoured growing areas, e.g. vineyards in France and Germany.

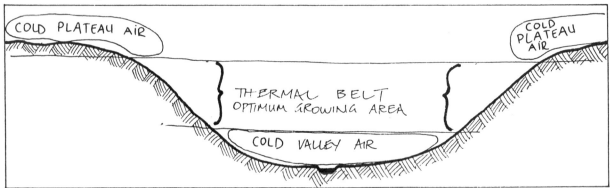

FIGURE 2.3 The "thermal belt" in a valley lies between layers of cold air and is the optimum area for house, orchard, and gardens.

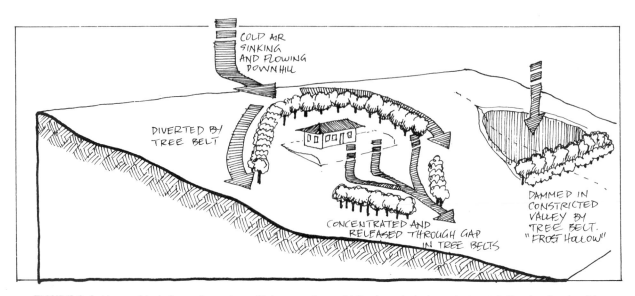

FIGURE 2.4 How cold air flows downslope. Note ways to avoid frost pockets by using vegetation to divert cold air.

However, this simple determination of frost works only on simple landscapes. Actual landscape, with its complex vegetational and topographical features, needs more observation and planning. Because cold air flows like treacle, rather than like water, it moves slowly around, over, and under sturdy objects, and is blocked by obstacles (buildings, trees, and landforms). For example, cold air flowing downhill towards the valley floor may be stopped by a forest above; in that case, the cold air is effectively dammed and will pool above the forest rather than in the valley. In order for the cold air to move downwards, large openings must be cut for air drainage (**Figure 2.4**); unless, in fact, the forest is protecting a house or vegetation immediately downslope.

Often, a constriction on the slope or near the valley floor will allow cold air to pool, and frosts may occur in any month (in temperate to cold climates). Houses sited above these constrictions will always be cold, whereas 20 metres away the perfect house site may well exist. Even in the subtropics, valleys below large bare plateaus can expect regular or occasional frosts following clear nights.

Winds

Although any site will be subject to global wind patterns or even catastrophic winds (cyclones and hurricanes), only the local prevailing winds matter when planning for microclimate. Topography can have quite an effect on local and regional persistent winds; in some mountain areas, regional prevailing winds may even come from the wrong direction because of a particular valley shape.

In valleys, slope winds are caused by rapid heating and cooling of the land on clear days and nights. Cooler air, being heavier, flows downhill. In a large valley system, small local winds follow a daily cycle (upslope and up-valley by day, downslope and down-valley at night).

Wind speeds increase on the windward side of ridges; decrease on the leeward side. (For any meaningful protection of the leeward side, however, wind speeds need to be at least 5 metres per second and the slopes 5° or more.) Wind speeds increase going uphill; decrease going downhill (**Figures 2.5a** and **2.5b**). And wind speed in-

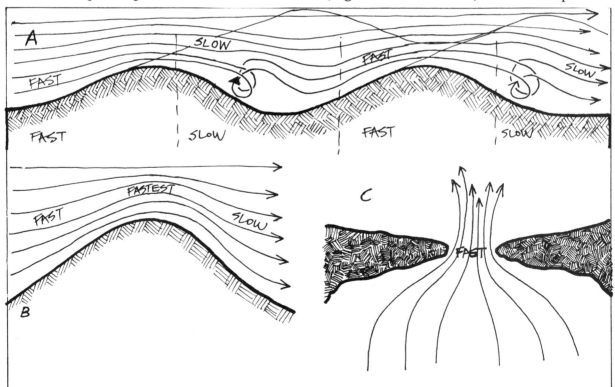

FIGURE 2.5 How wind behaves going up and down hills (A and B). In C, wind speeds increase at constrictions in landscape or vegetation.

creases going past a constriction (whether land-form or vegetation); this is called a "Venturi" effect (**Figure 2.5c**).

Near lakes or the sea, breezes play an important part in microclimate. Because of the marked temperature difference between large bodies of water and the land surface, air currents set up a cycle of offshore breezes. By day warm air rises over the land, allowing cool, heavier air from the sea to rush in. At night, as the land cools down, the process is reversed (**Figure 2.6**). In the tropics and subtropics these breezes bring welcome relief almost all year around, whereas in temperate regions they are more seasonal, usually appearing in summer. Houses, especially those in the tropics, are built to take advantage of the natural ventilation provided by these sea breezes. Conversely, in cool climates hedges are used to deflect these winds from house and garden.

We can tell from which direction the wind usually comes by examining the trees and bushes on the site. If they are bent in a particular direction, it means they are responding to frequent winds. At the sea-side, trees are almost flattened in response to strong breezes and salt spray coming off the ocean. If there is no vegetation on the site, stakes (1.5-1.8 metres tall) with cloth or plastic strips attached near the top can be driven into the ground at various places. Observing how often and in which direction the strips are blowing will tell us the usual wind direction. This method, of course, means observing the site throughout the year, so it is much better to analyse the surrounding vegetation, if possible.

Information on how to control winds with vegetation is given in a following section.

Elevation is also an important microclimatic factor. Temperatures decrease going uphill; every 100 metres (330 feet) of altitude is equivalent to 1° of latitude, so that at 1000 metres

FIGURE 2.6 How large bodies of water have an effect on coastal climate.

FIGURE 2.7 Effect of elevation on vegetation: cooler slopes even in tropical climates enable the growing of temperate species at higher elevations.

(3300 feet) on the equator, the temperatures are about equivalent to a climate 10° off the equator. This means that in a subtropical or tropical mountainous region, different vegetation can be grown. A typical planting sequence often found in the tropics from seashore to mountain is coconut, sugar cane, banana, tea, and pine (**Figure 2.7**), with each succeeding crop needing cooler conditions.

■ WATER MASSES

Large water masses such as the sea and big lakes warm up and cool off slowly, modifying the temperature of the surrounding area. In temperate climates, frost is rarely a problem near the sea, whereas just 20km inland frosts may occur during much of the winter.

Water also modifies temperature through evaporation. During evaporation, energy is drawn from the surrounding air, and as the temperature drops, the humidity increases. So even small lakes, pools, and ponds can be effective climate moderators, especially in arid areas. Fountains, for example, are found in many Mediterranean countries to provide evaporation and cooling of the courtyard.

Reflected light from water is also a consider-ation when designing a site. Although diffuse reflection from water surfaces is low, mirror reflection is usually high during the winter (when the sun is low in the sky). In the Main valley of Germany, reflected sunlight from the river is used to ripen grapes on steep hillsides. There-fore, the sun-facing banks or hills behind ponds, dams, lakes, and rivers can be considered favourable areas for marginal plants needing extra light and warmth. Houses situated on these banks or berms reap extra warmth (**Figure 2.8**).

■ STRUCTURES

Structures such as trellis, earth berms, green-house, fences, walls, and gazebos can affect microclimate on a small scale by modifying wind speed or temperature.

The *greenhouse* is the most useful structure for microclimate control in temperate regions, enabling almost any plant to be grown. Green-houses attached to the house are best for winter heating, saving winter fuel by day.

Earth berms or mounds affect microcli-mate in a variety of ways (**Figure 2.9**). They can:

• Block low sun on the west side, giving relief to house and garden towards evening.

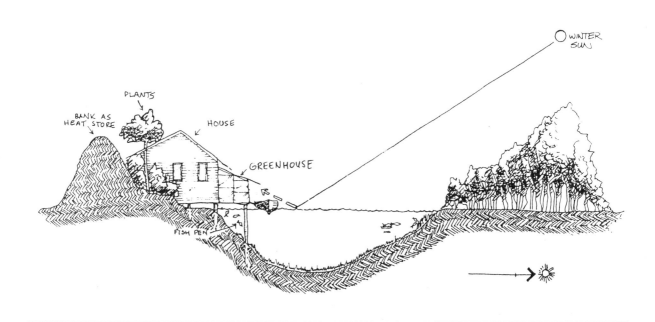

FIGURE 2.8 Ponds or dams as winter sun reflectors will increase dawn, evening, or winter heat on banks, benefiting greenhouse or vegetation (sun-facing banks are special ripening areas).

41

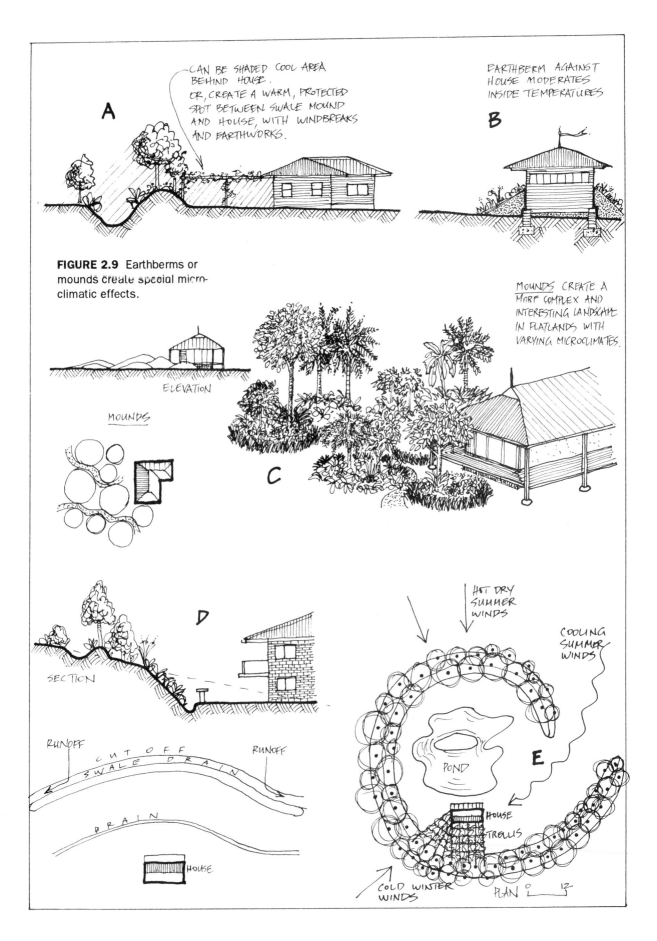

A — CAN BE SHADED COOL AREA BEHIND HOUSE. OR, CREATE A WARM, PROTECTED SPOT BETWEEN SWALE MOUND AND HOUSE, WITH WINDBREAKS AND EARTHWORKS.

B — EARTHBERM AGAINST HOUSE MODERATES INSIDE TEMPERATURES

FIGURE 2.9 Earthberms or mounds create special micro-climatic effects.

MOUNDS CREATE A MORE COMPLEX AND INTERESTING LANDSCAPE IN FLATLANDS WITH VARYING MICROCLIMATES.

ELEVATION

MOUNDS

C

D

SECTION

RUNOFF   CUT OFF   RUNOFF
SWALE   DRAIN

DRAIN

HOUSE

E

HOT DRY SUMMER WINDS

COOLING SUMMER WINDS

POND

HOUSE

TRELLIS

COLD WINTER WINDS

PLAN

- Block or channel winds.
- Offer insulation (soil retains heat and loses temperature gradually).
- Give privacy and block unpleasant views.
- Block traffic noises (sometimes by 80%); large earthberms between super highways and subdivisions are now commonplace.
- Provide more complex space for plants by increasing vertical space.

Sun-facing *walls* are also important in microclimate control. Like a sun-facing forest edge, walls provide shelter from winds and can be used to reflect winter sun. Dark stone walls absorb heat and re-radiate it at night, reducing frost risk. Plants placed in front of these will put on maximum growth. White-painted walls reflect heat (and so reduce heat gain); plants in front of these will ripen best. In Germany, experiments with tomatoes and peaches against both black and white walls showed more rapid plant growth against the black wall; yield, however, due to better ripening, was higher against a white wall.

Trellis is useful for quick wind protection, for dividing space around the house and garden, for making a microclimate (due to shading and heating), and as a temporary shelter over small trees to prevent sunburn.

Small structures around individual trees or plants create a microclimate of more moisture, less wind, and occasionally more heat. For trees, a variety of windbreaks are in use in various parts of the world: tyres, straw bales, old fertiliser bags, 44-gallon drums, etc. (**Figure 2.10**). In the garden, cold frames, cloches, and inverted plastic bottles can all be used to start plants in early spring.

■ SOILS

Soil has a small influence on microclimate due to the amount of heat it conducts and light it reflects, and also because of its differing water and air content.

As mulch conducts very little heat through to the soil, it is best to remove the mulch from growing areas in the spring so that the ground can warm up.

Mulch absorbs water readily and releases it

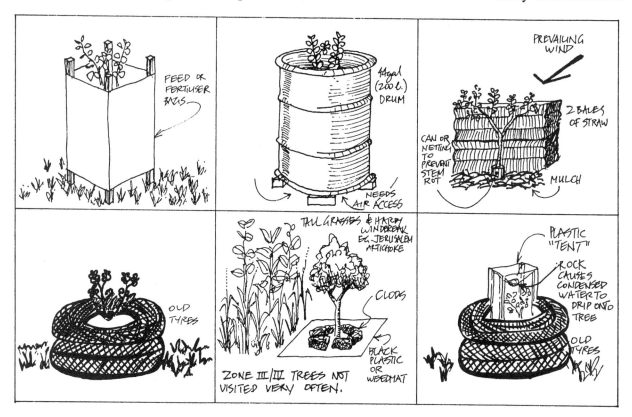

FIGURE 2.10 Climate-control strategies for individual (important) trees.

slowly to the soil, so it is an important aid in soil moisture retention during warm or windy periods.

■ VEGETATION

Vegetation has a profound effect on microclimate. It is the planting and use of vegetation (forest, woodland, windbreak, shrubs, and vines) that most shapes the microclimate of the site. Vegetation can modify the temperature of any particular site by:
- Transpiration
- Convective transfer of heat
- Shading
- Wind protection
- Insulation

Transpiration

Plants convert water in their leaves into water vapour which then passes from the leaf into the surrounding air. This process consumes energy, which causes the air around the plants to grow cooler (just like sweating in animals). As the temperature drops, humidity increases. For transpiration to work, water must be available. Many arid land cultures have techniques for making small areas, usually around the house, cooler. Canary Islanders use large earthenware pots filled with water and covered with hessian

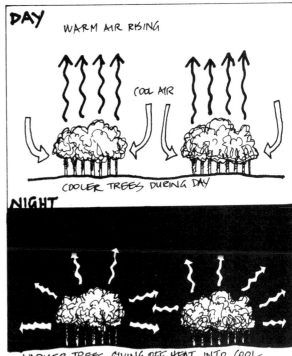

FIGURE 2.12 Forest is cooler during the day and warmer at night than surrounding air.

in small, plant-filled courtyards to cool the temperature of the surrounding rooms (**Figure 2.11**).

Convective Transfer of Heat

During the day plants absorb sun energy; in a forest or woodland large amounts of sun energy is absorbed by the leaf canopy, and the surrounding air is heated and rises. Cooler air is drawn into the forest, which remains cool during the day. At night this process is reversed, with warmer air than the ambient night temperature flowing out of the forest. The forest is insulated by its dense canopy of leaf cover, so that the flow is at the edges. Anyone who walks towards a forest at night will be able to feel the difference in air temperature (**Figure 2.12**).

Shade

Blocked sunlight has a powerful effect on microclimate. A piece of bare ground can cool down to 20% of its original temperature after the arrival of a shadow line from overhead foliage. Leaves have 3-6 times as much surface area for

FIGURE 2.11 Plant transpiration cools ambient temperature in hot-dry climates.

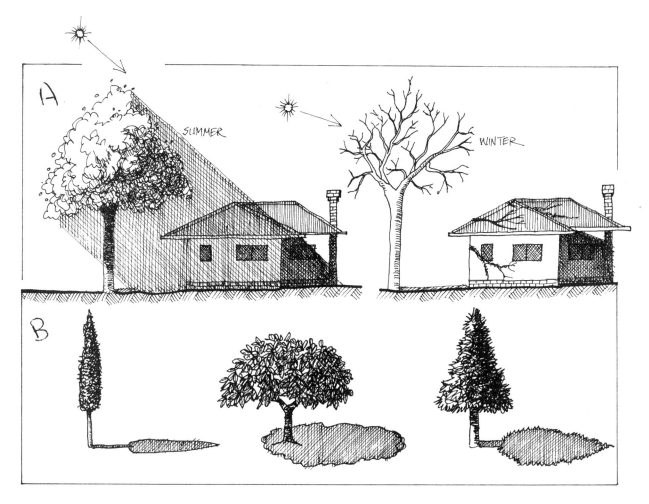

FIGURE 2.13 (A) Deciduous tree with seasonal shade effect on house. (B) Shape of shadow cast by different types of trees

energy interception as a canvas awning, depending on the density of the foliage. Trees with dense foliage can filter 75-90% of the sun's energy, while trees with loose foliage allow filtered sunlight. Also, trees with coarse or hairy leaves, and those with dark leaves, absorb sunlight and hence heat. Shiny, light-coloured plants reflect sunlight.

Designers can use this information to place appropriate plants in selected locations. For example, in climates where late afternoon sun is a problem, a dense hedge planted on the west side of a house not only provides shade, but also deflects westerly winds in winter. In contrast, a loose-foliage tree planted on the east or sun side of the house allows some protection from the sun in summer but lets in winter sun. Deciduous trees work in the same way as they drop their leaves in winter. The shape of a mature tree must

also be taken into account, be it round, oval, pyramidal, or columnar, as the shade it casts will also be of that particular shape (**Figure 2.13**).

To take advantage of sun reflection off shiny leaves, trees such as poplars can be planted in a parabolic arc around an orchard or house. With this arc facing the sun, the reflection off the shiny leaves will concentrate heat onto a point within the arc, making this area drier and warmer (**Figure 2.14**). Such suntraps also work on a slope, as the vegetation will trap warm air rising up the hill. This shape allows cold air coming downhill to flow around it, minimising frost danger, and, depending on wind direction, helps to deflect cold winds around buildings or fields.

Wind

Windbreaks have been used for years to shelter houses, animals, and crops from wind,

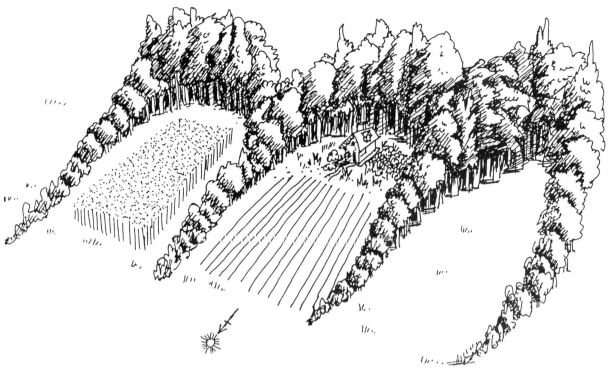

FIGURE 2.14 Sun trap shapes for house and fields

and are the most effective in microclimate control. The benefits of windbreaks are as follows: They

- Reduce wind velocity and soil erosion.
- Protect wind-sensitive plants such as kiwifruit.
- Reduce crop losses caused by the shaking out of seed or grains.
- Modify air and soil temperatures (the soil can be as much as 10°F higher in a sheltered area).
- Increase available moisture due to dew formation on the leaves of trees.
- Reduce the number of animal deaths during cold storms.
- Reduce animal stress caused by summer heat.
- Reduce feed requirements if animals can browse some of the windbreak trees (honey locust, carob).
- Provide timber and fencing materials for the farm (from thinned or old windbreak).
- Enhance the habitat of insect-eating birds.
- Improve living and working conditions around the house and farm.
- Provide sources of nectar for bees and improve conditions for pollination of crop (less wind to deflect bees).

Windbreak shape depends largely on the crop, site, and climatic condition. **Figure 2.15** shows a variety of windbreak groupings.

Dense and permeable windbreaks are used for different purposes. Dense windbreaks give the greatest protection leeward at 2-5 times the height of the trees (**Figure 2.15c**). However, the protection falls off rapidly because negative pressure forms downwind and draws the wind back down. The pressure difference also dries out the soil. On the other hand, a permeable windbreak (**Figure 2.15d**) allows air to flow through, and although the initial protection is not as great as with a dense windbreak, the protection continues for a greater distance (25-30 times the height). **Figure 2.16, 7-9** shows other effective windbreaks for intensive cropping, while **Figure 2.16, 1-6** illustrates some ineffective windbreaks.

As shelterbelts can be used to serve other useful functions, consider the inherent characteristics of particular trees. Almost any tree

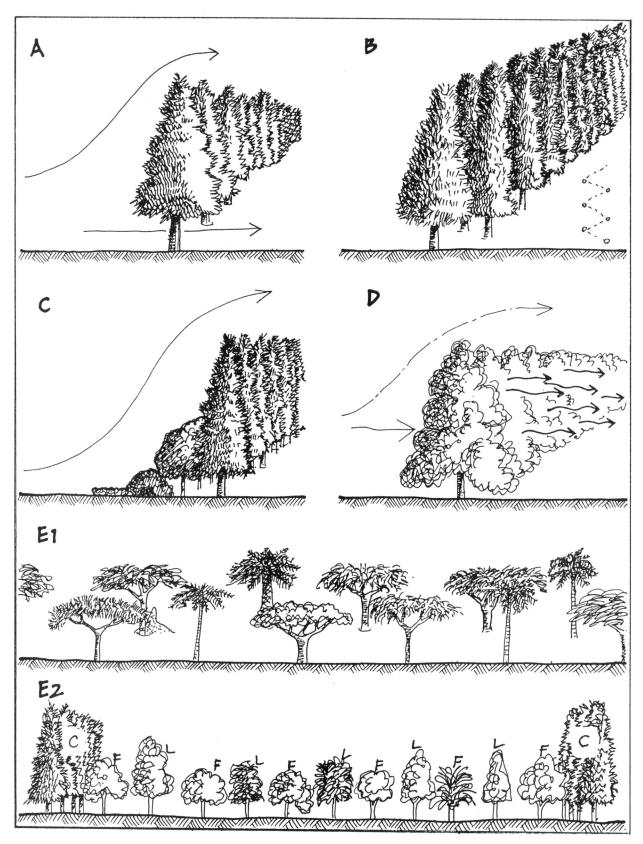

FIGURE 2.15 Windbreak configurations.
There is no "best" windbreak; every crop, site, or condition needs specific design.
Here, (A) suits ridges, (B) tall vine crop, (C) coasts, (D) fields, (E1) desert crops, (E2) temperate orchards (L=legume, F=fruit tree, C=conifer or windbreak).

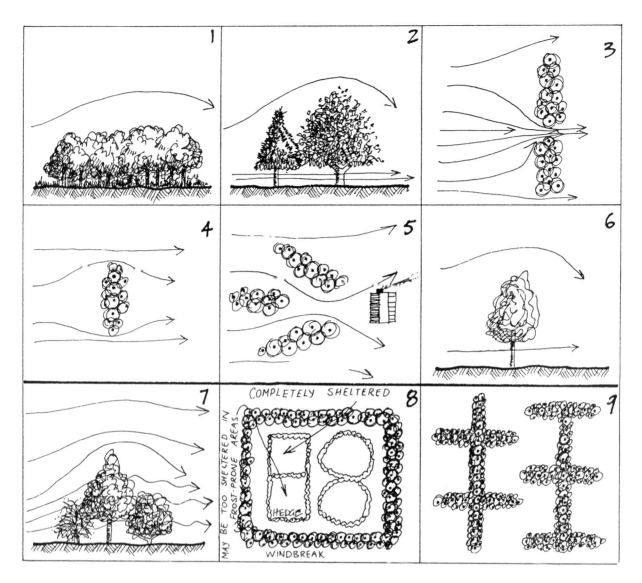

FIGURE 2.16 INEFFECTIVE WINDBREAKS: (1) too wide (2) lower branches cut, grazed off, or dead: wind acceleration (3) velocity in gaps (Venturi effect) (4) too short (5) not sited perpendicular to winds (6) only one row of trees.
EFFECTIVE WINDBREAKS: (7) small, medium and large tree rows with approximately 50% permeability for best airflow (8) combining windbreak and hedges for total shelter where appropriate (coast sea winds, desert dessicating winds). Not appropriate in frost-prone areas unless windbreaks are broken for cold air flow release (9) T-shaped windbreaks for regular air flow and protection.

provides wind protection (as long as it is not itself wind-sensitive), privacy, and shelter for animals. What else can it do? Some species (leguminous trees, alders) fix nitrogen in the soil, can be coppiced for firewood (willow, eucalypts), provide foliage for animal fodder (*Coprosma repens*, leucaena, willow), provide honey (eucalyptus, leatherwood), yield nuts for animal or human food (oak, chestnut), act as a fire retardant (*Coprosma repens*, black wattle), and are useful in erosion control (strong-rooted trees such as willow and poplar).

The negative aspects of trees must also be taken into account. Some have vigorous root systems that compete for water and nutrients with crops or pasture next to them. We can either accept this in exchange for the benefits they provide, or maintain a yearly deep ripping to partially cut back the root system, reducing competition.

Windbreaks are quickly started using fast-growing shrubs and trees interplanted with slower-growing (but longer-lived) trees. As these trees (usually hardwoods) are slowly growing, fast-growing trees provide nectar for bees, forage for animals, and mulch for the garden, and are later harvested for firewood or polewood. Note that trees used for windbreaks will not yield much fruit (as the wind shakes them off), and should not be relied on for commercial use.

Coastlines present particular difficulties. Across the great unmodifiable plain of water, winds arrive at gale force, carrying salt and abrasive sand grains. To protect ourselves from these winds we choose vegetation with:

• rough bark, like palms (able to withstand sandblast);

• hard, needle-like leaves, such as hardy coastal pines (*Araucaria*), tamarisks, casuarina (to resist severe dessication); or

• fleshy leaves, like the ice-plant, taupata, agaves, and *Euphorbias* (which retain moisture).

The best guide, when choosing species, is to observe successful species already growing in the local area. **Figure 2.17** shows a possible seaside planting sequence.

Insulation

Bushes and vines planted next to a building protect it from wind and also add a pocket of insulating air between the building and the vegetation, thus protecting it against heat loss.

Snow is also a good insulator for buildings if it is piled up on the roof or against the shade-side wall, thus reducing heating costs. Shrubs and trees help to trap snow in preferred areas. Snow beneath windbreaks insulates the ground, ensuring an even temperature (acting in much the same way as deep litter, or mulch). Snow thaws slowly on sunny days, ensuring a slow warming of the ground. Depending on what is planted near the shelterbelt, this can have a negative or positive effect. Spring bulbs will bloom later than bulbs planted in fast-thawing environments.

Special Vegetation Strategies

Plants in the form of vines, ground covers, and shrubs are very useful in microclimate control.

Vines and Trellis

In very windy areas, plants suffer most from lack of wind shelter. The fastest possible aid in these cases is to build trellis at near right angles to the house walls. Such trellis has a multiple effect: it separates living space into a recreational, garden, or service area; prevents the flow of cold winds along walls (and acts as a sun trap); and itself presents a basic structure for

FIGURE 2.17 Sample seaside planting sequence.

vine crop. Trellis structures may curve out from the house corners, or simply break up a facade on institutional buildings (schools, prisons), offering several places for benches, lawns and gardens.

Frequently, large buildings and roads converge to make wind tunnels. Large boulders, trees and shrubs, and trellis convert them to sheltered and sinuous access, and block dust, cold, and noise as a side effect. This is true of all driveways, service roads, blind entries and minor trafficways.

Besides their windbreak potential, vines are fast-growing (4.5-6 metres a year in hot, wet climates) and can be used for fast shade while trees are growing. Be careful in selecting the right species for the climate and situation, as vines can be rampant and difficult to eradicate once established. Pruning may be an option in these cases. Some vines grow into mortar, wood shingles, window frames, and downpipes and gutters, so it is best to find out the characteristics of any particular vine before designing it in.

Vines have good insulating properties if placed over roofs and on walls. Thick vines can reduce heat gain by 70% and heat loss by 30%. Ivy in temperate regions have been used for hundreds of years to insulate brick buildings in both summer and winter. Deciduous vines such as grape, wisteria and Virginia creeper can be placed on the sun-side of houses or gardens in temperate or hot, arid regions for shading.

Ground covers and mulch

Bare ground is much hotter and colder seasonally than protected ground. Ground is best uncovered in the spring when new plantings are underway and the soil needs warming; otherwise earth is best covered with mulch and living groundcovers. Natural groundcovers (grass, creeping plants) and mulch have the following features: They

• Reduce heat build-up by evaporating water and shading the ground.

• Do not re-radiate heat (as plastics and pavements do).

• Protect the ground from erosion.

• Do not reflect light, so can be used to reduce glare.

• Keep the ground warm or cool, depending on the weather.

• Act as a weed barrier (although occasional weeding may be necessary).

Non-grass groundcovers are planted beneath trees (young fruit trees grow poorly in grass) as "living mulch". Depending upon climate, these can be dichondra, *Dolichos*, lupin, and mass plantings of marigold. If the groundcover is also a vine, it may have to be chopped back every now and then. A locally-occurring or native legume is most useful for nitrogen-fixing.

Shrubs

Shrubs provide a moisture envelope around a tree, and can give frost protection in marginal areas. Miriam and Jim Tyler in a marginal area of New Zealand planted tagasaste 0.6-0.9 metres from avocadoes to protect young trees from frost. The tagasaste was slashed 2-3 times during the summer for firewood and mulch around the tree, and eventually cut out altogether.

Shrubs make good garden dividers and are used for wind protection, especially in seaside gardens. Suitable species must be chosen to eliminate time spent in trimming and dealing with roots.

Shrubs and even existing "noxious weeds" used as nurse vegetation provide mulch, shade, nitrogen fixation, and protection from frost, wind, and animals. On the north coast of New Zealand Ian Robertson planted a commercial tamarillo crop directly into slashed gorse, while Dick Nicholls has developed a sequence for establishing native forest in country taken over by gorse. Both are using this already-present weed for its positive qualities (mulch, soil amelioration, frost protection), slashing it over a four-year period around a nucleus planting of trees. Gradually the trees shade out the gorse. The same can be done in large patches of blackberries.

In permaculture soils are not considered to be a severe limiting factor. The soil ecology, over some years and with the proper attention, can be changed and improved. The house site and Zone I is not selected purely on the basis of soils, although if good soils exist on a particular area and most other factors make it a good location, by all means set the house and gardens there to save a year or two.

Very few soils are totally worthless; there are always colonizing or pioneer species to start with. Almonds and olives do well on rocky areas with very little soil; blackcurrants and butternuts grow on poorly-drained sites; blueberries thrive in very acid soils; and honey locust can grow on the most alkaline of soils.

On any site, a basic soil survey is necessary to find out the pH (for garden and orchard), drainage capacity, and types of vegetation already growing. From there we can decide the species we need to plant and the type of soil improvement we need to make, depending on the scale of land use. Obviously, the greatest effort will go into the home garden and orchard, while outlying areas will receive broadscale attention.

Bare soil is damaged soil, and occurs only where people or introduced animals have interfered with the natural ecological balance. Once soil has been bared, it is easily damaged by sun, wind, and water. Cultivating the soil then not only damages soil life processes, but may even cause more extensive soil losses.

The three main approaches to minimal soil loss in permaculture, which aerate and add nutrients to the soil, are:
• Growing forests and shrubberies to protect the soil (afforestation).
• Using ploughs that do not turn the soil (soil conditioning).
• Encouraging life forms, especially worms, to aerate compacted soils (mulching or composting).

The first two deal with large areas, the last with small areas. Forestry and soil conditioning produce their own mulch, whereas mulch can be applied to small gardens.

Often, the pest plants of which we complain (lantana, capeweed, blackberry, mullein, thistle and so on) are an indication that soil damage has occurred. Some of these plants are pioneers, and will eventually modify the soil so that other species are able to grow.

The mark of a good soil is an adequate level of soil moisture, oxygen, nutrient, and organic matter. Soils are formed and nourished by the cyclic process of plant roots drawing water and mineral nutrients from the subsoil, and the fall of leaves, fruit, and other detritus to the ground.

The steps for soil rehabilitation include:
• Preventing erosion by covering all exposed soil, afforesting potential erosion areas (such as steep slopes, gullies, creek banks, and road embankments), and controlling overland water flow (by use of swales, diversion drains or chisel plow). Use local fast-growing plant species. Logs can be placed across slope to catch silt and water, and plants placed behind them.
• Adding organic matter to the soil. Broadscale: cover crops, green manure crops. Small scale: kitchen scraps, dead vegetation.
• Loosening compacted earth and providing air to the soil. Broadscale: chisel plow and soil reconditioning machinery. Small scale: loosening with a fork.
• Modifying the pH, or growing plants suitable for specific pH areas (more economical than changing the pH). For acid soils, chalk and limestone, gypsum, magnesite, and dolomite are all used to slowly raise the pH. For alkaline soils, use acidic phosphate, and urine for potash. For all soils, blood and bone, manures, and compost help to bring the pH toward neutral.
• Correcting nutrient deficiencies with organic minerals (e.g. manganese, phosphorus, potassium) and animal manures, green manures. Pelleted seeds and foliar sprays are economical ways to add nutrients to plants.
• Encouraging biological activity; earthworms and other soil organisms indicate a

healthy soil.

In general, soils can be created or rehabilitated by the following methods:
- Management of plants and animals.
- Mechanical conditioning (broadscale).
- Building a soil (garden scale).

■ MANAGEMENT OF PLANTS AND ANIMALS

Managing livestock to minimise compaction and over-grazing is part of the skill of soil building and preservation. On severely-eroded land, livestock may have to be totally excluded. Some farmers introduce earthworms to their pastures, and sow deep-rooted plants (daikon radish, chicory) to break up and aerate soils. Daikon radish, tree or shrub legumes, earthworms, root associates for plants (rhizobia) all aerate, supply soil nutrient, or build soil by leaf fall and root action.

Mulch, cover crops, and green manure crops prevent erosion, add organic matter and nutrients to the soil, buffer soil from extremes of heat and cold, and protect soil water from evaporation.

There are two categories of mulch: "dead", which is dried out, decayed, or dying (straw, dried leaves, recently-cut vegetation); and "living", which grows underneath trees and shrubs. Dead mulch must be collected (sometimes from scattered locations), while living mulch needs management (sowing and cutting back, sometimes re-seeding).

Cover crops are those planted to protect the soil after a main crop has been harvested. In temperate climates, these are usually planted in winter and include rye, vetch, clover, buckwheat, lupin, barley, oats, etc. which can either be harvested or turned into the soil to increase organic matter.

Green manures are grown specifically for soil improvement, and are usually leguminous, supplying both carbon and nitrogen to the soil (cowpea, clover, field peas, lupins, vetch, *Dolichos*). Legume crops are mulched, or turned into the soil before the plants mature, to take advantage of the nitrogen release from the roots as the plant dies (if allowed to flower and set seed, most of the nitrogen is used up).

■ BROADSCALE SOIL RECONDITIONING

Australia, Europe and the United States now manufacture chisel ploughs that aerate and loosen large acreages of soil. A circular coulter slits the ground (which must be neither too dry nor too wet), and the slit is followed by a steel shank and subterranean shoe which opens the ground below the surface to form an air pocket *without turning over the soil* (**Figure 2.18**). Rather, it is gently lifted. Rain penetrates and is absorbed; soil temperatures rise, roots grow and die to make humus, and the country comes to life again.

There is no point in going more than 10cm in the first treatment, and to 15-22cm in subsequent treatments. The roots of plants, nourished by warmth and air, will then penetrate to 30cm in pasture, more in forests.

Seed can be dropped in the thin furrows; legumes sown in this way produces a green manure crop or a bumper harvest. No fertiliser or top dressing is needed, only the beneficial effect of entrapped air beneath the earth, and the follow-up work of soil life and plant roots on the re-opened soil. However, on severely degraded soils, an initial top-dressing of phosphate or grossly-deficient trace elements could be used.

Once the soil is on the way back to health, tree and field crops can be planted. A season spent on bringing the soil to life is not a season wasted, for trees respond more vigorously to the new soil conditions, and make up for lost time: an olive or carob struggling to survive in the original condition of compacted soils will make 90cm to 1.2 metres growth in improved soil, and may well bear in 3 or 4 years instead of 10-15 years.

There is only one rule in the pattern of this sort of "ploughing", and that is to drive the tractor and plough slightly downhill from the valley across slope to the ridges, making herringbones of the land. The slit channels, many hundreds of them, thus become the easiest way for water to move. Because the surface is little

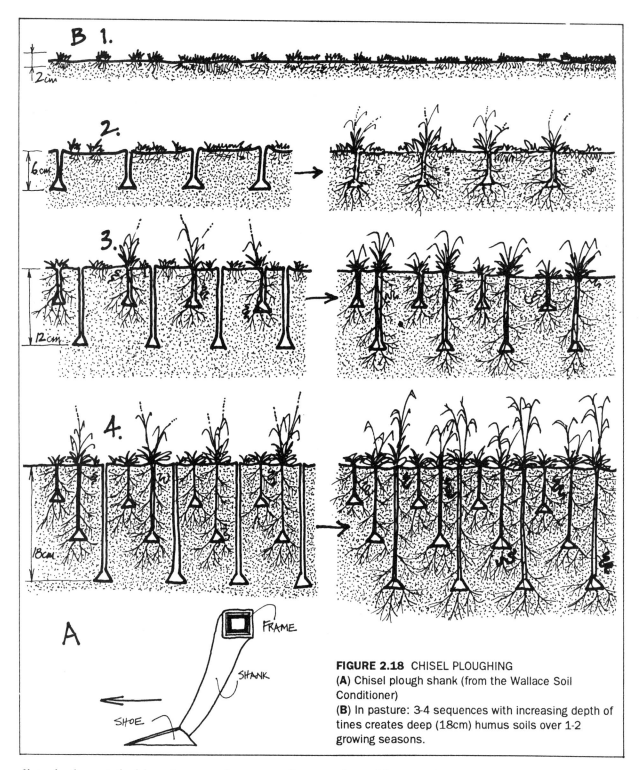

FIGURE 2.18 CHISEL PLOUGHING
(**A**) Chisel plough shank (from the Wallace Soil Conditioner)
(**B**) In pasture: 3-4 sequences with increasing depth of tines creates deep (18cm) humus soils over 1-2 growing seasons.

disturbed, roots hold against erosion even after fresh "ploughing", water soaks in and life processes are speeded up.

To summarise, the results of soil rehabilitation are as follows:

• Living soils: earth worms add alkaline manure and act as living plungers, sucking down air and hence nitrogen.

• Friable and open soil through which water penetrates easily as weak carbonic and humic acid, freeing soil elements for plants, and buffering pH changes.

• Aerated soil, which stays warmer in winter and cooler in summer.

• Absorbent soil, preventing runoff and rapid evaporation to the air. Plant material soaks up night moisture for later use.

• Dead roots as plant and animal food, making more air spaces and tunnels in the soil, and fixing nitrogen as part of their decomposition cycle.

• Easy root penetration of new plantings, whether these are annual or perennial crops.

• Permanent change in the soil, if it is not again trodden, rolled, pounded, ploughed or degraded by chemicals into lifelessness.

What soil conditioners achieve, Fukuoka does with deep-rooted plants such as daikon radish and lucerne, but his system has not been compacted by heavy machinery or domestic stock. Even strong roots often cannot break up hard pans.

■ BUILDING A GARDEN SOIL

Gardeners normally build soil by a combination of three processes:

• Raise or lower beds (shape the earth) to aid water retention or drainage, and sometimes carefully level the bed surface for effective flood irrigation;

• Mix compost or humic materials in the soil, and also supply clay, sand, or nutrient to bring it to balance; and

• Mulch to reduce water loss and sun effect, or erosion.

Gardeners can, by these methods, create soils anywhere. Allied techniques involve growing such compost or manurial "tea" materials as hedgerow, herbs, or soft-leaf plants (e.g. comfrey) as plots or rows within or around the garden, and by the use of trellis, shadecloth (or palm fronds), greenhouse, and trickle irrigation to regulate wind, light, or heat effect.

Mulching should be recognised as one of the larger initial costs in developing a permaculture. Although materials such as seaweed, bean and grain husks, spoiled hay and animal manures are very cheap (or free), transportation and application can be costly, usually as labour. This is because of the great bulk of such materials. For example, 15 cubic metres of sawdust does not go far when sheet mulching. Chipping machines—as used by councils to dispose of tree prunings—would be useful for direct mulching, using the scrub vegetation, tree tops and bark from land clearing and timber felling.

■ SPECIAL CLIMATIC CONSIDERATIONS
Tropical Soils

In the tropics, as elsewhere, bare-soil cultivation is not sustainable. Wet terraces and ponds will sustain production if they form 15% or so of the total landscape, but for areas over 1 hectare we must plant borders, hedges and woodlots, and intercrop with woody legumes. About 80-85% of all plant nutrients are held in the *vegetation* in tropical areas, and crops therefore cannot be sustainable without the nutrient from tree leaf-drop and root mass. Soil organisms will build up only after shrubs and trees are established.

Soils cleared of vegetation are likely to need calcium, silica, and such easily-leached nutrients as sulphur, potash and nitrogen. Initially, phosphates (as bird manures or rock dusts) may also have to be added. Try some cement dust, or use bamboo or grain husk mulches in gardens for adding calcium and silica. For nitrogen and potash, plant leguminous trees and add their leaves to soils, if necessary via livestock as forage and manure. Restrict agricultural crops to 20% of the total plant cover, preferably as strips in forest systems; this should build soils and prevent nutrient loss. Even grasslands need large legume trees at 20-30 metre spacing (or 20-40 per hectare) to sustain production. Above all, keep slopes of 15° or more *terraced or wooded* to prevent soil loss and to avoid severe erosion.

Dryland Soils

The major characteristic of arid soils is that of alkalinity (pH 8.0-10.5) caused by calcium, magnesium, or alkaline salts (carbonates) evaporated from surface soils. Thus, we are most likely to find that trace minerals (zinc, copper, iron) are hardly available, so that deficiency symptoms show up in both plants and people.

Once we analyse the soil for such deficiencies, however, we can supply them to plants as foliar sprays, and to the earth as compost and mulch.

In drylands, soil humus can rapidly decompose (in dry, cracked soils) to nitrates with heat and water, giving a sometimes lethal flush of nitrate to new seedlings. Mulches or litter on top of the soils and tree roots prevents both soil cracking and the effect of rapid temperature gains that cook feeder roots at the surface.

In home gardens, soils can be treated on a small scale. Where free-draining or non-wetting sand is a problem, bentonite (a volcanic fine clay which swells up and holds water) is of great help in flood-irrigated beds. Conversely, where clay is causing problems with water absorption, adding gypsum lets water penetrate further into the clay particles. Where salty soils or salty waters are a problem, the garden beds must be mounded up or raised, so salt can leach down out of the growing bed onto the paths.

2.6
WATER

Available water affects the type of permaculture for a site, and depends on the following:
- distribution and reliability of local rainfall;
- drainage and water retention properties of the soil;
- soil cover (vegetation, mulches); animals (stocking densities, species); and
- plants (species, requirements).

Although the first factor is fixed, the other three can be controlled.

A priority on any property is to identify water sources and reserve sites for water storages (dams, tanks). Wherever possible, use the slope benefits (or raise tanks) to give gravity flow to use points.

Fitting species to specific sites reduces watering needs. For example, olives and almonds on dry hillsides require no water (besides rain) once established.

Water storages for growing fish and plants are usually very differently designed structures than those for stock watering or irrigation alone. For instance, many small ponds are better suited to fish culture than very large storages. Graded bottoms of from 75cm to 2 metres depth suit many fish, while storage ponds for water need to be 3-6 metres deep to be worthwhile on large acreages.

■ WATER COLLECTION AND DISPERSAL
We can get water from rain runoff (surface or underground), springs (groundwater seepage), and permanent or intermittent streams. To bring this water to storage areas, we use diversion channels (sealed or otherwise impermeable), pipes leading from springs, and roofs or any other sealed surface collecting direct rainfall.

Diversion channels are gently sloping drains used to lead water away from valleys and streams and into storages and irrigation systems, or into sand beds or swales for absorption (**Figure 2.19** and **2.20**). They are built to *flow* after rain, and can be constructed so that the overflow of one dam enters the feeder channel of the next.

Direct rainfall can be captured by large roof areas, sealed roads, or even sealed hillsides in

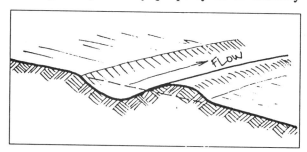

FIGURE 2.19 Diversion drains flow from streams to dams, or collect water flow and carry it to dams. They are a vital part of any rainwater harvesting system.

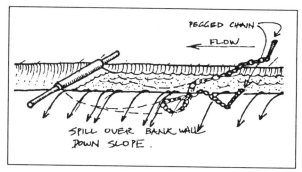

FIGURE 2.20 A plastic sheet, one end supported by the channel banks, the other weighted by a chain, forms a temporary dam, causing the channel to flood and irrigate land downhill.

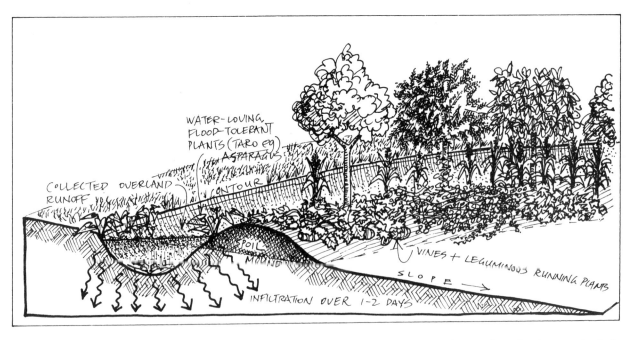

FIGURE 2.21 Swales on contour do not flow; they first stop and then absorb overland water flow. Swales are planted with trees or shrubs on the mound side.

arid areas leading to water tanks.

■ SWALES

Water absorption into the ground is usually achieved through soil conditioning, and by swales. Swales are long, level excavations, which can vary greatly in width and treatment from small ridges in gardens, rock-piles thrown across slope, or deliberately-excavated hollows in flatlands and low-slope landscapes (**Figure 2.21**).

Like soil conditioning or soil loosening systems, swales are intended to store water in the underlying soils or sediments. They work to intercept all overland water flow, to hold it for a few hours or days, and to let it slowly infiltrate as groundwater recharge into soils and tree root systems. Trees are the essential components of swale planting systems, and *must* accompany swaling, especially in arid areas (to reduce salt buildup).

Swales are built on contour or on dead level survey lines, as they are not intended to allow water to flow. Their function is just to hold water, thus, the base is ripped, gravelled, sanded, loosened, or dressed with gypsum to allow wa-

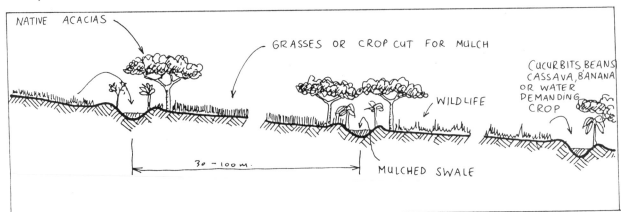

FIGURE 2.22 Swale distance in arid lands is greater than in humid climates. Swales on slopes grow fodder legumes and hardy trees. The interswale space can be sown to grasses or grains after rains on chisel ploughed soils.

ter infiltration. The earth spoil is normally mounded downhill or (in flat areas) spread. Water enters from roads, roof areas, tank overflows, greywater systems, or diversion channels.

The distance between swales can be from 3 to 20 times the average swale width (depending on rainfall). Given a swale base of 1-2 metres, the interswale space (area between swales) should be 3-18 metres. In the former case (3 metres), rainfall would exceed 127cm (50 inches), and in the latter it would be 25cm (10 inches) or less. In humid areas, the interswale is fully planted with hardy or mulch-producing species. In very dry areas, it may be fairly bare and exist mainly to run water into swales, with most of the vegetation planted on the banks (**Figure 2.22**).

After an initial series of rains that soak in a metre or more, trees are seeded or planted on either bank or side slopes of the swales. This can take two wet periods. It may take about 3-10 years for tree belts to shade the swale base, and

to start humus accumulation from leaves. Early in the life of an unplanted swale, water absorption can be slow, but the efficiency of absorption increases with age due to tree root and humus effects.

Swales are used in arid lands to collect silt, to recharge groundwater and to prevent rapid erosion; and in humid lands mainly to retard erosion. In all cases, they also serve as planting areas.

■ TANKS AND DAMS

Most usable water is stored in tanks and dams. Tanks are made from rolled galvanised iron; concrete; ferro-cement; wood; or clay (rendered), and can accept water from roof runoff; runoff over a sealed surface into a silt trap (if necessary); or pumped from a dam.

The minor problems associated with tanks are easily solved. For mosquitoes, *Gambusia* or other type of small, larvae-eating fish are introduced, or the tank completely screened and covered. The inlet is screened to exclude leaves,

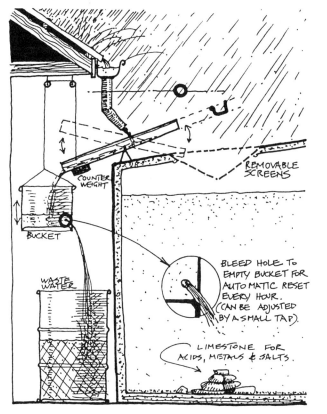
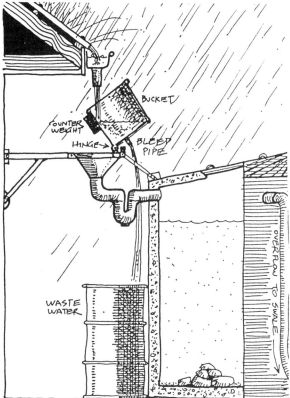

FIGURE 2.23 Two ways to reject the first water flow off a roof (for dirt and debris). Both these systems automatically reset when empty.

etc. from the roof or sealed ground (**Figure 2.23**). Some people object to algae on the sides and bottom of the tank; however, this velvety film is composed of life organisms, filtrating and purifying the water. The water outlet pipe should be at least 6cm from the bottom of the tank so as not to disturb the algae.

Small dams and earth tanks have two primary uses. The minor use is to provide watering points for range animals, wildlife, and domestic stock. The second and major use is to store surplus runoff water for use over dry periods for domestic use or irrigation. These need to be carefully designed with respect to such factors as safety, water harvesting, total landscape layout, outlet systems, and placement relative to the usage areas (preferably providing gravity flow).

Open-water storages are most appropriate in

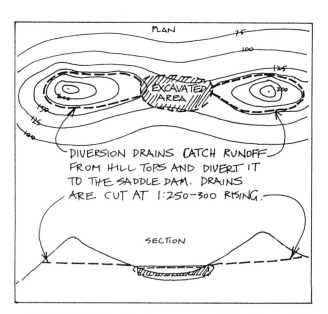

FIGURE 2.24 Saddle dams are useful for fire control, wildlife, limited irrigation. It is the highest type of dam in the landscape that fills from hill runoff.

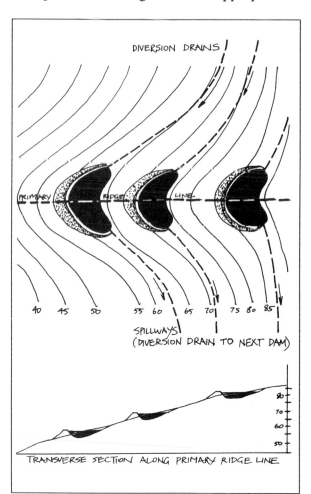

FIGURE 2.25 Ridgepoint dams are built on plateau areas of ridges.

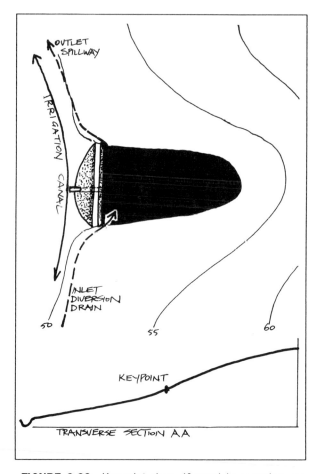

FIGURE 2.26 Keypoint dam. If used in a series, no spillway is built and the overflow goes to the next dam, and eventually to a stream. Fitted with an irrigation system for lower slopes.

humid areas. There is a danger that similar storages created in arid to sub-humid areas will have negative effects, as evaporation from open water storages inevitably concentrates dissolved salts.

The following are common dam types and their uses in humid landscapes:

Saddle dams are usually the highest available storages, on saddles or hollows in the skyline profile of hills. Saddle dams can be fully excavated below ground (grade) or welled on either side of, or both sides of, the saddle (**Figure 2.24**). Uses are for wildlife, stock, and high storage.

Ridgepoint dams or "horseshoe" dams are built on the sub-plateaus of flattened ridges, usually on a descending ridgeline, and below saddle dams. The shape is typically of a horse's hoof. They can be made below grade, or walled by earth banks (**Figure 2.25**). Uses are as for saddle dams.

Keypoint dams are located in the valleys of secondary, or minor, streams. They are built at the highest practical construction point in the hill profile; this place can be judged by eye, and a descending contour will then pick up all other keypoints on the main valley (**Figure 2.26**). Uses are primarily to store irrigation water. Note that a second or third series can be run below this primary series of dams to larger barrier dams, and that the spillway of the last dam in a series can be run along the contour to meet the main valley, effectively spilling surplus to streams (**Figure 2.27**). *Barrier dams* are built across a flowing or intermittent streambed and therefore need ample spillways and very careful construction.

Contour dam walls can be built on contour wherever the slope is 8% or less, or sufficiently flat. Contours (and dam walls can be concave or

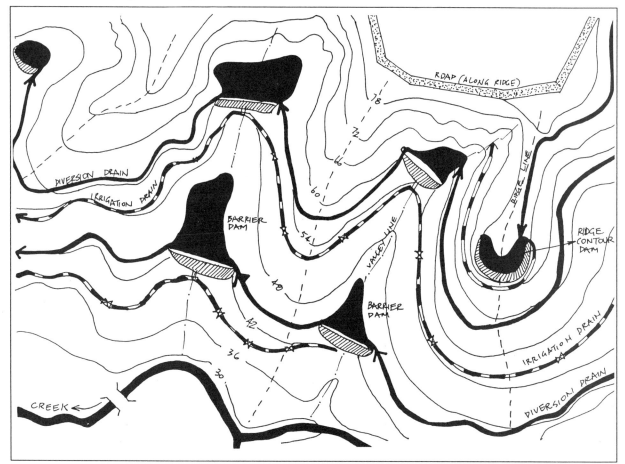

FIGURE 2.27 P.A. Yeomans' "Keyline" system provides drought-proofing for farms with very low maintenance and operating costs; his books cover total water design for foothill farms, access, tree belts, soil creation, low tillage, and creative water storage.

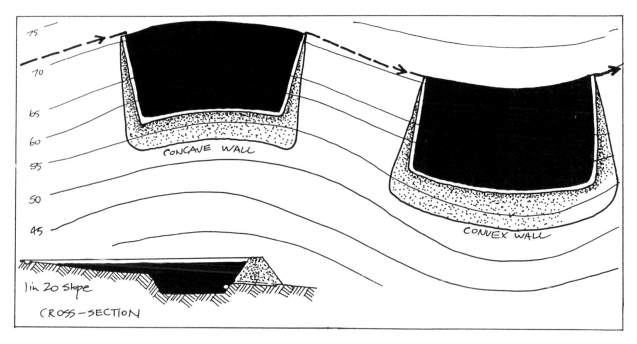

FIGURE 2.28 Contour dams are practical on slopes of 8° or less, as part of a general series of dams on farms.

convex to the fall line across the slope. Uses are for irrigation, aquaculture, or flood-flow basins in semi-arid areas (**Figure 2.28**).

■ WATER DIVERSION AND STORAGE IN DRYLANDS

In most dryland areas of the world, groundwaters and aquifers are over-drawn, and agricultures and cities depending on such temporary events are doomed to failure. It is sad indeed that instead of being used to grow a sustainable tree crop and forest system, these precious aquifers and groundwaters go mostly to produce an annual export crop of grains or grain legumes.

Thin sheets of runoff water, which generally appear after 1-2cm of rain, can be led cross-slope to storages. These diversion drains are made of earth, stone, concrete, or piped to storages, or end in artificial hollows and basins dug to receive them. As a general rule, such growing basins, terraces, or pits are built to harvest a runoff area about 20 times their own area (8-10 hectares of runoff area is led to 0.4 of a hectare of trees or seasonal crop).

Native or adapted trees are the best use of such sites, but at times of good rains, grains,

melons, or vegetable crops can be grown on an opportunistic basis.

When we concentrate overland water flow, especially in the fragile desert environment, we must also allow a safe overflow or outlet of excessive rains, or we risk creating gullies. Where we can grow grasses, a grassed and fenced downhill spillway will resist erosion; or we can build a carefully-laid stone spillway down very steep slopes or stepped terraces.

Every dryland situation, given some study of water movement, sand movement, and some data on infiltration and runoff, can be shaped to make a growing site. If regenerated areas are protected from browsing and exploitation, such useful trees as figs, mulberries, pistachio nuts, and acacias will persist and even spread.

2.7
SITING IMPORTANT INFRASTRUCTURE

The boundaries of the property have been walked during the observation and research stages, and many favourable niches and resources have been discovered. We can now look at other factors involved in siting such important infrastructure as access, the house, and fences.

■ ACCESS

Access to the house site and around the property is important for establishing and maintaining the site. During the first few years, materials are continually brought in to build up infrastructure.

Depending on the type of transportation used (car, four-wheel drive, tractor, wheelbarrow), roads, tracks, and paths will have to be sited, made, and maintained. Access should be sited in such a way that it will need little upkeep, as a misplaced road will cost more in time and money than almost anything else. Although the layout design will vary according to climate, landform, and available resources, a few principles are as follows:

1. Roads should run along contours, with no steep slopes and with good drainage to reduce erosion. If possible, they are sited, in hill country, on the centre of a ridge, so that water can easily drain off. Roads can be built in valleys, but will require more upkeep, especially in high-rainfall areas.

2. Roads should, wherever possible, fulfil other functions, such as dam walls and firebreaks. The road as a water-collector can also be considered, with runoff led to swales and dams, or pooled and used as a silt trap for potting mix or tree mulch (**Figure 2.29**).

3. On hilly sites, a high road or tractor access should be established to give access to all areas from above (it's easier to move materials downslope than up).

4. Smaller roads and foot tracks are worked out to complement access roads in an integrated layout plan worked out early in the design process.

Water drainage is the most important aspect of building a road. The road should be shaped to

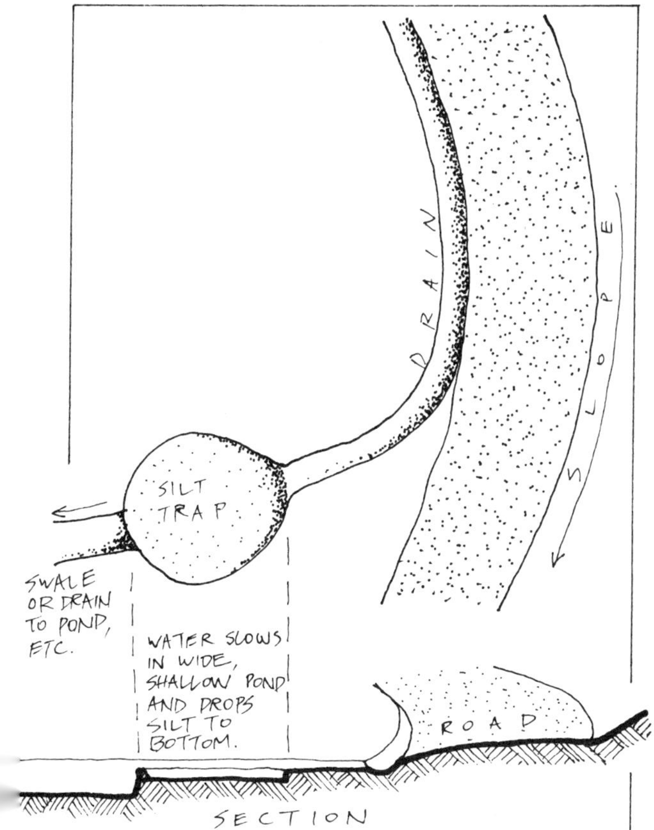

FIGURE 2.29 Road runoff led to silt trap; silt dug out periodically for potting soil.

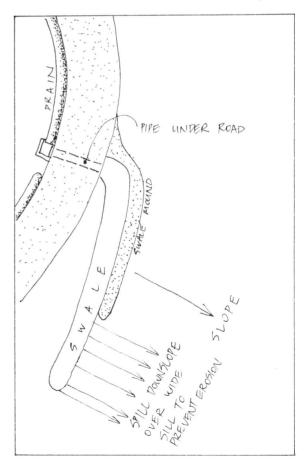

FIGURE 2.30 Water from hill roads led into under-road pipe and then to swales to prevent soil erosion.

accommodate drains and their outlets. If water cannot be drained off on the same side as the ditch or culvert on the inside of the road, it needs to be piped under the road (**Figure 2.30**) to a drain led either to a stream or other area where erosion will not occur (dam, diversion channel, swale).

Always finish the driveway upgrade to the house, no matter if you have to drop it a little in order to run it upgrade to the house. There are several reasons for this: most driveways that descend to the house carry water down around the house area, making it hard to drain properly. Also, when your car battery is flat, you'll be able to get a gravity roll. In a snowy climate, it is wise to have a road in the sun for a faster melt; the same is true in a particularly wet climate when roads are muddy and slippery.

■ SITING THE HOUSE

Although house siting varies with climate, there are certain rules to follow and errors to avoid.

The closer to a main road, the better. Long roads to a housesite are costly, difficult to maintain, and give rise to a sense of isolation.

In climates where house heating is needed, choose the aspect which receives the sun, especially in winter. In tropical or equatorial areas, any aspect will do, but the house is oriented to receive cooling breezes rather than direct sun.

Do not build on any slope above 14° or below 2-3° (for reasonable drainage). Mid-way up a gentle slope is best in order to avoid frost and to receive cooling breezes.

Site the house so that its water source is upslope for gravity-feed. Also make sure that waste products (sewage, greywater) are not discharged where they will pollute streams or groundwaters. Use trees or vegetation buffers as nutrient sponges.

Build close to power supplies, whether they are public supply or water, solar, or wind power. It is very expensive to pipe energy from source to house as there is a loss of power in transmission (for alternative energies) and costly poles and cable to lay (public sources). For village needs, use communal sources of energy to save money.

Use the site landforms or existing vegetation to shelter from damaging winds, or locate the house to take advantage of cooling breezes. Strong wind sites enable the use of wind power.

Don't build the house on the best soils. Also, check the subsoil for drainage (test by digging a hole a metre deep and fill with water; within a minute there should be a visible lowering of that water).

Consider both current and future privacy needs; to avoid noise and exhaust pollution, houses should be built off the main highway. Privacy is achieved by vegetation, but to lessen traffic noise, large earthbanks must be built between the road and the house.

Although most of us put "view" as a priority, it can lead to siting a house inappropriately, usually on a hilltop where access is difficult and winds are frequent. So we may have to sacrifice the view from the house, and instead build a little retreat up on the hill, with comfortable chairs. You can go along with your guests across Zone II and into Zone III for the panoramic effect, and stay at home for the close-in view. You can have bird-attracting shrubs right next to the window, or a large fish and duck pond with an island or two nearby where there's always something moving around, always something to look at.

Sometimes you can build upwards, and look at your view from a cupola on the roof. A retired sea captain might have a house with a bridge deck up above, so that the view to the sea is always clear. He'll have a telescope on the bridge deck. When storms come, he goes up to his wheelhouse and gets out onto the open deck. He's up there making sure no rocks come up in the middle of the night!

The most common errors in house siting are:
• Building at the top of an exposed ridge or hill. Winds can come from any direction, and the house is at risk from fire (fire speeds intensify going uphill). Water must be pumped, adding to

energy costs (the major energy cost will be in heating and cooling the house).

• Locating a house in the bush, setting up a conflict between the forest (and its inhabitants) and yourself for light, nutrient, and space. Vegetation must be cleared for the house, garden, and orchard.

• Building on riverflats or low gullies (prone to flooding or waterlogging); steep, unstable land (slips, mud flows, avalanches); infilled land (subsidence); near active volcanoes; near rising sea levels (due to global warming); or in fact anywhere inevitable disaster threatens.

■ FENCING

Fencing and enclosures are essential and priorities should be decided early in the planning stages. A general boundary can be established first, to keep out stock and wildlife. Total control of animals (especially small wildlife such as possums and rabbits) is not possible on a large scale, and should be confined to Zone I. From this sturdy, small-mesh inner fence, other fences can be built as needed, perhaps eventually enclosing Zone II (with large mesh or even barbed wire, thorny trees/shrubs, or electric fence). Fencing priorities might include a chicken run and orchard.

Instead of wire fence, unpalatable hedge species can be grown over time. A dense, spiny hedgerow with a low stone wall is virtually impenetrable to most animals, and is used throughout the world where wire is expensive or hard to get. Fences, ditches, stone walls, and hedgerows should function not only as enclosures or protection from stock, but have other uses. Fences serve as trellis, and stone walls as special ripening areas. Hedgerows provide fruit, nuts, animal forage, bee forage, bird habitat, and wood products (bamboo). In temperate climates, a mixed hedgerow of tagasaste (fast-growing, provides seeds for chickens, bee forage, and shelter), hawthorn (slow-growing, tough and spiny, provides berries, forage for bees and nesting sites for small birds), and hazel (thicket-forming and impenetrable, provides nuts) is much more useful than a single-species hedge.

Different plants such as *Prosopis*, *Euphorbias*, and spiny acacias do the same work in tropical and desert regions.

■ DECIDING PRIORITIES

Once access and house site have been decided upon, the design can become more complex and focus on the built-up area and its surroundings. This is when zones, sectors and slope can be analysed in a broad sense (saving detail for later), and even at this point the house location could change as a result of these investigations.

Sectors are then sketched in as areas defining wind direction, aspect, good and bad views, flood or fire-prone areas, and water flow direction. Zones are sketched out on a ground plan, with Zone 0 marking the house and Zones I-V marking increasingly distant or difficult-access areas.

Once we have broadly placed our elements by zones, sectors, elevation, and function, we go further into the design process by considering specific plant and animal species.

The plan should be drawn up to be taken in stages, to break up the job into easily-achieved parts. Important components are placed in those stages that are needed early in development, which might include: access roads, water provision, fences or hedges, energy systems, windbreaks, house and garden, and plant nursery. Secondary priorities might include fire control, erosion control, and soil rehabilitation.

So many plant species and individuals of each species are needed in the first 2-6 years that a small plant nursery should be established to supply the 4,000-10,000 plants that might be placed on a hectare. While these are growing in their pots and tubes, we can fence and prepare the soil, lay in the water system, and then plant them out to a carefully-designed, long-term plan.

Provision for future energy conservation systems must be left open, so that the whole site is marked out for wind, tide, water, or sun systems. Even if these cannot be implemented in the first few years, the space is reserved under annual crop or short term use.

When it comes to implementation, the first structures and designs should be those that generate energy; second, those which save energy; and only finally, those which consume energy.

Applying such criteria, many questions will answer themselves, for example:

Where should I build my greenhouse?

On consideration of energy alone:

• First, against dwellings as heat sources and storages, and to grow food.

• Second, against non-dwelling structures, as heat sources.

• Thirdly, as part of animal housing, with heat, manure and gas exchange.

• And only finally, or perhaps never, as free-standing, all-glazed structures.

How should I deal with wind which prevents my growing on site?

• First, by planting any tree or shrub, useful or not (wormwood, pampas, pine, taupata) that is cheap or free locally, grows very quickly, can be grown from large rooted cuttings or divisions, and that will *survive*.

• Second, by structures, especially trellis, loose or dry-stone wall, ditch, bank, and small hedgerow throughout the garden.

• Third, by broadscale cutting or seedling plantings of hardy species.

• And lastly, by useful permanent hedge planted under the protection of the above strategies.

What is worth main-cropping?

Only a few plant species are worth extensive main cropping. Ignoring the commercial value for the moment, there are three main considerations:

1. main crop which needs little attention after establishment (potatoes, corn, pumpkin, hardy fruits and vines);

2. and which is easy to harvest, store and use;

3. also, may form a staple in the diet (potatoes, taro, cassava, corn, pumpkin, nuts, and high energy-value fruits).

Commercially, we should also consider crops of:

4. high economic value, even if they are difficult to harvest (berries, cherries, crocus for saffron);

5. or hard to keep (melons, peaches, papaya);

6. or rare but in wide demand (ginseng, spices, teas, dyes, oils);

7. or particularly suited to site (sugar maple, cider gum, pistachios, water chestnut, cranberry, cactus).

The designer should be ever alert to local features, microclimates, and needs, endeavouring to turn what is already in place to advantage, rather than to bring in new structures, and hence, new energy.

2.8
DESIGN FOR CATASTROPHE

Every area in the world has the potential for such catastrophic events as fires, floods, droughts, earthquakes, volcanoes, or hurricanes. The best we can do is to design the site with such events in mind so that we lessen damage to property and loss of life.

■ FIRE

Fire is the most common catastrophe, occurring in dry, windy periods after forest litter build-up. Fire intensity depends on the fuel quantity, type, and distribution, wind speed and direction, and general topography (fire travels fast uphill, so ridges are most likely to be severely burned). The greatest danger is *radiant heat* from the fire front, which quickly kills plants and animals.

Fire usually comes from a specific direction (varying according to location and topography), so that there is generally only one fire sector to be concerned about. However, fire can come from any direction, so it is best to protect the most valuable elements of the system first (buildings, animal pens, machinery, and orchards).

Strategies for dealing with fire include:

• **Reduce the fuel** in the fire sector by (a) managing the forest floor (clearing litter, cutting dead logs for firewood), (b) mowing or using short grazers (geese, wallaby) to keep grass short, and (c) using non-fuel surfaces, such as roads, ponds and dams, sheet mulch, or green crop, between the fire sector and the house.

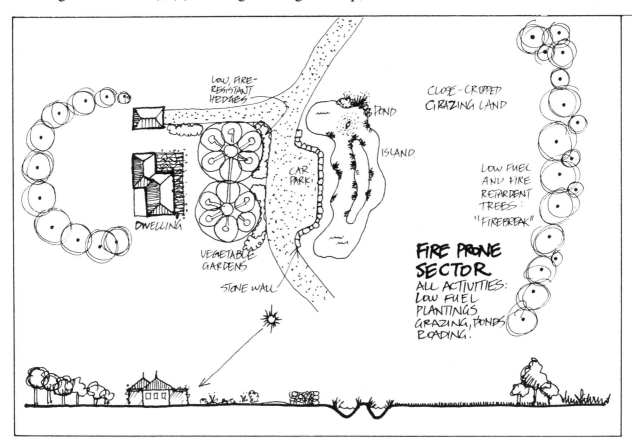

FIGURE 2.31 Combined design for fire safety: radiation shields, multiple downhill firebreak, selected "wet" plants, and fuel reduction near the house.

FIGURE 2.32 Fire defence planting (with livestock) for small towns or house sites.

• **Create fire shadows** to reduce the effects of radiant heat by (a) non-fuel structures (ponds, earthbanks, stone walls), and (b) plantings of fire-retardant species such as lilies, coprosmas, willows (which may be killed, but will slow down the fire). **Figure 2.31**.

• **Plant a windbreak** of fire-retardant species to reduce wind during a fire (**Figure 2.32**)

As the house is usually the most difficult and expensive part of the site to replace, it is important to plan for house safety by providing:

• A brick or concrete apron (to 1 metre) around the house, with doormats removed;

• Metal screens on the windows;

• Corrugated iron or fire-resistant roof;

• Large sprinklers on the roof and around the house, and at least one hour's worth of water at a source easily brought to the house (fire burns through unburied plastic waterpipes, and electric pumps may fail);

• Tennis balls to plug up the downpipes of rain gutters (which can then be filled with water).

Fire-resistant plants for the fire sector are those that combine the following features: (a) a high water content, (b) high ash content, (c) little mulch or litter drop, or fast-decomposing litter, (d) are evergreen and (e) are fleshy or sappy.

Some fire-resistant trees are: figs, willows, mulberries, *Coprosma*, *Monstera*, and some of the acacias (*Acacia dealbata*, *A. decurrens*, *A. saligna*, *A. sophorae*, *A. baileyana*, among others).

Some fire-resistant ground covers include passionfruit, ivy, comfrey, taro, various succulents, wormwood, *Dichondra repens*, aloe and agave species, iceplant, sweet potato, wandering jew, onionweed, sunflowers and pumpkins.

■ EARTHQUAKE, FLOODS, AND HURRICANES

In earthquake-prone areas, build houses of materials that bend or breathe (bamboo, ferrocement, wood). During an earthquake, escape into a clump of bamboo; it has a strong structural root mat which is difficult to tear apart.

For floods, look up the flood periodicity and height records, allow a large margin of safety, and do not site houses on floodplains. Steep slopes that have been cleared of vegetation are death-traps during severe rains, as mudflows accelerate rapidly downhill.

In hurricane- or cyclone-prone areas, build out of flexible materials, and make the shape of the house roof sharply angled at 45° so that windforce pushes the building down. Plant a bamboo windbreak (it bends with the wind), and consider a survival garden in a sheltered place. Many Pacific Islanders have such gardens, made up of important plant stock, in a sheltered part of the island, so that gardens can be replanted after everything else blows away.

2.9 REFERENCES AND FURTHER READING

Geiger, Rudolf, *The Climate Near the Ground*, Harvard University Press, New York, 1950.

Chang, Jen-Hu, *Climate and Agriculture*, Aldine Pub. Co., Chicago, 1968.

Cox, George W. and Michael D. Atkins, *Agricultural Ecology*, W.H. Freeman & Co., San Francisco, 1979.

Daubenmire, Rexford F., *Plants and Environment*, Wiley International, 1974.

Fukuoka, Masanobu, *The One-Straw Revolution*, Rodale Press, Emmaus, PA, 1978.
Out of print - available from libraries

Howard, Sir Albert, *An Agricultural Testament*, Oxford University Press, 1943.

Moffat, Anne Simon & Marc Schiler, *Landscape Design That Saves Energy*, William Morrow & Co., New York, 1981.

Nelson, Kenneth D., *Design and Construction of Small Earth Dams*, Inkata Press, Melb., Australia, 1985.

Yeomans, P.A., *Water for Every Farm/Using the Keyline Plan*,   Phone (075) 916 281 Queensland, Australia.

CHAPTER 3

Pattern Understanding

3.1 INTRODUCTION

While elevation drawings and contour maps can be used to depict various components of a landscape, they fail to depict the dynamic or living quality of a site. "The map is not the territory" (Bateson 1972).

In a natural landscape, each element is part of the greater whole, a sophisticated and intricate web of connections and energy flows. If we attempt to create landscapes using a strictly objective viewpoint, we will produce awkward and dysfunctional designs because all living systems are more than just a sum of their parts. Our culture has tried to define the landscape scientifically, by collecting extensive data about its parts.

These methods are much like the group of blind mullahs in the Sufi tale, who try to describe an elephant.

"I see" said the first, grasping a leg, "an elephant is like a tree".

"I see", said the second, holding the tail, "an elephant is like a snake"

Another, feeling the ear, said "an elephant is surely much like a thick carpet".

Traditional societies have used patterns to effectively understand and interact with their landscape - they do not differentiate between themselves and their environment but see the elements as relatives. Thus all traditional knowledge and science was recorded in the form of motifs or patterns as carvings, weaving, stone and earth constructions and tattoos. Every motif was accompanied by songs or stories that told of its meaning, and song was reinforced by sacred dances to ensure 'muscle' memory of the stories. The important records were of history (sagas), creation myths, genealogies of ancestors, navigation, and cyclic phenomena such as tides, weather, star cycles, and crops or wild harvest seasons. Everybody in tribal societies had access to a good part of this knowledge, including the names and uses of important plants. Many intact tribes still maintain this knowledge.

After writing was invented, patterned knowledge was neglected, and modern systems work entirely with alphabet and number, with symbols, books, or electronic storage of data. Much of human society cannot access, and none can accurately remember, knowledge stored in these ways. That is, patterned and rhythmic knowledge is unforgettable; symbolic knowledge is unmemorable.

All through this book, we evolve patterns of ground planning, as in all design, and all the parts of any design have to be fitted into a common-sense template or pattern. To understand the basic pattern into which all natural systems fit, we will dissect a tree, and try to

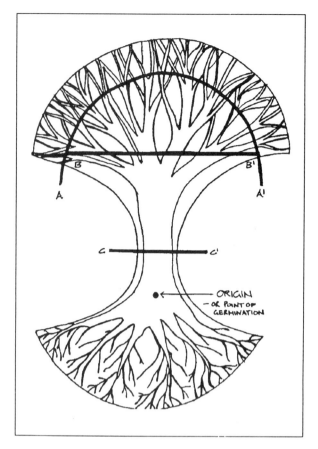

FIGURE 3.1 General Core Model

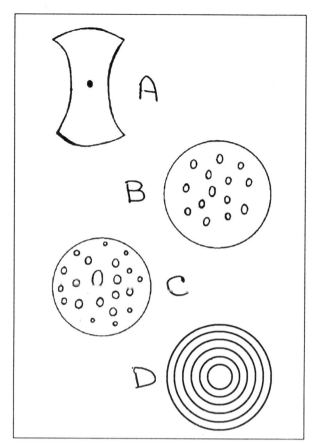

FIGURE 3.2 Sections of General Core Model.

make sense of the rules of flow (sap movement) and form, of growth and expansion. We will use the form of a tree, which is typical of all natural phenomena. **(Figure 3.1)**

PATTERN IN NATURE

The essential lines of a tree can be imposed on a real tree, and form a sort of double-headed axe. **(Figure 3.2A)** This is the motif of older European tribes - the "woman's symbol". If we cut the tree off the line A-A[1] **(Figure 3.2B)** we see the branch stubs in plan, not unlike the scatter of limpets on a stone; each branch section is of roughly even diameter. But if we cut across the tree on the line B-B[1] **(Figure 3.2C)** we get another graded pattern, which is like a scatter of lichen on a stone, the oldest at centre, the smallest and youngest on the outside. A cross-cut of the trunk, C-C[1] **(Figure 3.2D)** gives us a classic

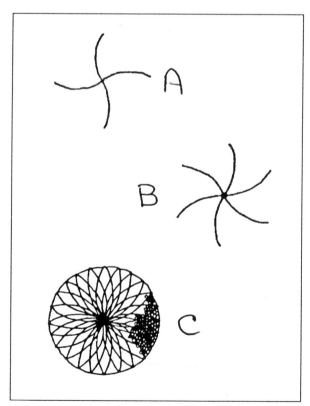

FIGURE 3.3 Sap flow wirls

68

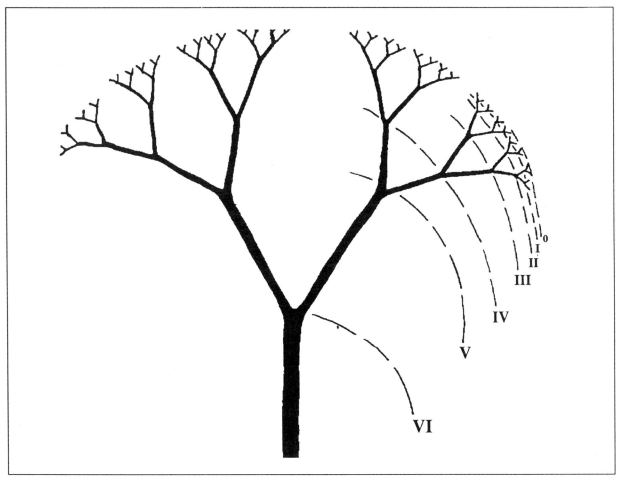

FIGURE 3.4 Dendritic or tree-like pattern typical of lightening,mineral crystals, blood vessels etc,.

target pattern which is an annular record of growth seasons, found in shell fish and fish scales. We would get such a pattern in a nest of bowls. As Latin for nest is Nidus, we call this pattern *Annidated* or nested (each in the other). In effect, the whole tree is an annidation of younger trees, grown-over year after year.

As we know that the tree spirals out of the ground, the branching from above describes a whorl; or more specifically, the flow path of a molecule of sap whirls. **(Figure 3.3A)** Sap flow in the stems is on the outside (the xylem cells), and the roots a whorl in the opposite sense. **(Figure 3.3B)** Sap flow in the roots is along the core (the phloem cells). If we combine branch and root whorls, **(Figure 3.3C)** we get the twin overlaid spirals we find in all leaves, flower petals, sunflower seed-heads, pine-cones, and pineapples, and obviously at the point where the seed germinated, the origin, we get a weave of

cells changing from inner to outer, left spin to right spin or plus to minus.

The tree branches 5-8 times, as do rivers, and the number of branches arising from each larger stem average 3, while each is about 2 times longer than the next smallest. The angle between each branch is about 36-38º. **(Figure 3.4).** This form is typical of lightning, mineral crystals, blood vessels etc., which follow roughly the same rules. Such patterns are called tree-like or dendritic.

The Roman numerals I to V are called the Orders of branching, and rarely exceed 7 in all; each represents a count at larger or longer sizes. This number of orders of size is common to a very large range of phenomena, which can be assembled into bins or clumps of sizes, e.g. for settlements, we name such bins, cities, towns, villages, hamlets. Also clouds, mountains, celestial bodies, and dunes, waves etc., etc., etc.;

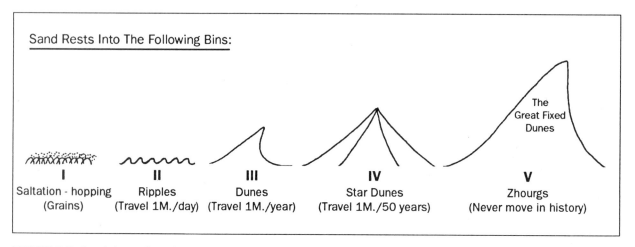

Sand Rests Into The Following Bins:

I	II	III	IV	V
Saltation - hopping (Grains)	Ripples (Travel 1M./day)	Dunes (Travel 1M./year)	Star Dunes (Travel 1M./50 years)	Zhourgs (Never move in history)

FIGURE 3.5 Sand dunes form five orders of size common to a very large range of phenomena.

all have a limited set of sizes, as do tree branches. We can speak of cascades of size, or quanta, and this means that most sizes go into specific bins, and there are few (if any) sizes between. For instance, sand dunes set into 5 orders as illustrated in **Figure 3.5**.

Thus everything in nature (cats, kangaroo, whirlpools, whirlwinds, roads, etc., etc.,) occur in a very few sizes, and the speed of movement in each size differs. Large things move slowly due to greater inertia, small things fast, and very small things again move slowly because of viscosity. Orders are limited in their sizes at the larger scale, by sheer mass and at the smaller scale, by molecular forces. It is becoming clear that the patterns in a single tree form represent all the patterns found in nature. Even the bark of many trees exhibit furrows like a web of cells or an elongated honeycomb net.

To return to the general tree form as a whole, we see that the axe illustration (**Figure 3.2A**) is the simple form. A set of such forms creates a vertebra, or a skeleton, (**Figure 3.6A**) and it fits together, as if tiled. Latin for tile is Tessera, and we call tiled surfaces *Tessellated*. A cloud form contains a torus or doughnut, (**Figure 3.6B**) and various "single paths" of a molecule in the model, show traditional motifs. (**Figure 3.6C**)

A general sense of form is emerging, and we can see many such forms in nature, thus gaining understanding of function, and a clear

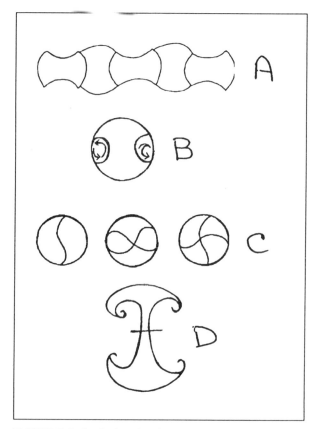

FIGURE 3.6 Analysing the characteristics, needs,

insight into how to design in order with nature. Many natural 'curled' forms exist in explosions or fast-growing fungi. The shapes (**Figure 3.6D**) are called 'Overbeck jets' and occur in fluids, often in complex folds. They also appear as motifs, e.g. in Maori tattooing as stylised fern shoots. You will see them when rivers flood into the sea, in lava flows, and when the sea attacks the land.

70

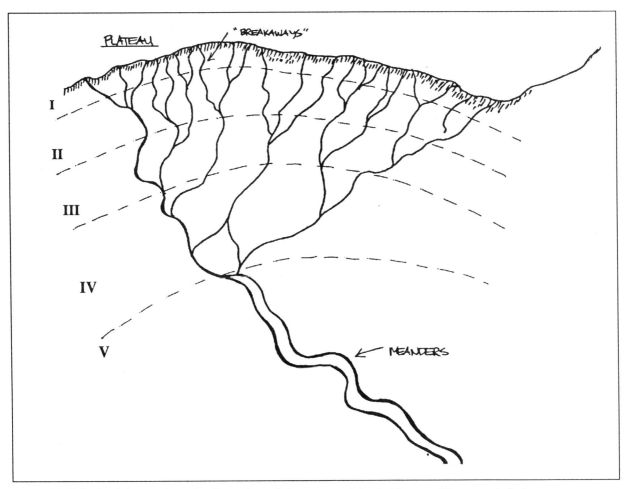

FIGURE 3.7 Dryland river system showing orders of streams, each having its own differences in factor.

We can build all natural forms from parts of a tree and such forms are called "self-similar"; seashells demonstrate the same geometric spirals as trees. It was the study of these natural forms, and their meanings, that gave rise to pattern literacy in tribal peoples; a simple pattern which can contain vast knowledge.

3.3
PATTERN IN DESIGN

Good designers try to fit all their components into a pleasing and functional form, to obey the rules of flow and order, and to compact space. A well-designed house allows airflow to heat and cool its mass, and a village or town road system based on natural branching will have no gridlock of traffic.

The herb spiral (illustrated on Page 96) is a good example of applying pattern. All the basic culinary herbs can be planted in an ascending spiral of earth on a 2m wide base, ascending to 1m high. All the herbs are accessible, there are variable aspects and good drainage, and the spiral can be watered with just one sprinkler.

Using our observations of plant guilds, or harmonious assemblies, we can design forests which mimic natural systems but use climatically-adapted food plants.

Moreover, as the diagram of a dryland river (**Figure 3.7**) shows, not only run-off for water harvesting, but species, sediments, and flow, all vary according to the order of the streams; once you absorb this sort of information, you can much more easily plan for stability and production in landscapes. (**Figure 3.8**)

As this river illustrates, all life forms, all sediments, all run-off in storms of 12mm (1/2 inch) or more varies with the order of the streams.

71

Factor	<Orders>				
	I	II	III	IV	V
Sediment:	Angular Stones	Angular Shingle	Gravels	Coarse Sands	Deep Sands
Vegetation:	Hard Dark Shrubs	Taller shrubs	Occasional Trees	Large Light Tree	Great Trees & Vines
Runoff: (% of total)	86-90%	55-65%	40-50%	30-35% ("average")	8-15%

FIGURE 3.8 Forming a chart according to Orders and Factors assists in more easily absorbing information and planning for stability and production in landscapes.

If we know where we are in the order, we know not only what vegetation to expect, but what to plant; not only what run-off the expect, but the spacing of swales or water harvesting systems. Most desert villages are located in orders III or IV where run-off is ample, some good soils and fresh minerals occur, and where the distance between streams is large enough to allow for fields, but not so large as to incur drought.

If we observe the meanderings of rivers, in order IV and V, we find that large, leafy, often white-barked trees (white gums, sycamores) grow on the 'inside' curves where the river deposits sand; and dark trees with thick fissured bark (alligator juniper, casuarinas, ironbark gums) grow on the outside curves, above the cliffs where the river is cutting in to the landscape. At each curve, these tree groups swap over as does the cliff or sediments, so we have a yang-yin-yang-yin effect.

Close observation will show you where rodents and reptiles burrow, where baboons live, where pecaries forage for fruit and roots, and (for the birds) in which order of rivers and branches in trees they belong. Fish, of course, are often highly adapted to one order of flow-speed, or spawn in one order and live in another.

In designing with nature, rather than against it, we can create landscapes that operate like healthy natural systems, where energy is conserved, wastes are recycled and resources made abundant.

3.4
REFERENCES AND FURTHER READING

Alexander, Christopher et al, A Pattern Language, Oxford University Press, 1977. Instances successful design strategies for towns, buildings

Mollison, B., Permaculture - A Designers' Manual, Tagari Publications 1988

Murphy, Tim and Kevin Dahl, Patterning: A Theory of Natural Design Landscape Ecology, Conference paper 1990.

Thompson, D'arcy W., On Growth and Form, Cambridge University Press, 1952. Mutiple examples of forms in nature, spirals.

CHAPTER 4

Structures

INTRODUCTION

Efficient house design is based on the natural energies coming into the system (sun, wind, rain), on surrounding vegetation, and on commonsense building practices. Many houses are already built, or being built, without any thought of future oil shortages and present rising fuel costs. However, with correct house placement and design for the climate, simple technological aids such as solar hot water heaters, and perhaps some adjustment in behaviour (so that we choose warmer clothing or open and shut vents leading to an attached greenhouse), we can reduce or eliminate our dependence on fossil fuel energy for heating and cooling the house.

General rules for siting the house, and planning the surrounding vegetation for microclimate control are discussed in Chapter 2, which should be read in conjunction with this chapter.

■ THE HOUSE AS A WORKSPACE

Houses have become more fully occupied places, especially with the modern trend towards using the house as a workplace. It is cheaper to adapt the house to a small manufacturing and office area than to buy or rent these facilities separately (and is especially cheaper in transport costs). Some home industries and occupations are: cabinet-making; pottery; small seed company; honey production; desktop publishing (magazines, newsletters, books); preserves, pickles; accounting, computer, and secretarial services; medical and psychotherapy; advertising, photography, and real estate services.

The living/working areas may need careful thought and re-design. Bedrooms, for example, are converted to office, computer, or studio space by elevating beds to sit on low chests of drawers or by raising the ceiling and building the bed into a small, warm alcove above the office. Space-saving design involves the same kind of "stacking" found in nature, where shelves, elevated beds, and ceiling or roof structures mimic herb layer species, understorey, and plant canopy.

■ HOUSE AND GARDEN INTERGRATION

Just as there is no reason to strictly separate garden from farm, so house and garden are very much integrated. Turf roofs, and wall and trellis vines added to the house provide external insulation, and greenhouses and shadehouses produce food and climate modification. One of the most pleasant summer views I know is that from Elizabeth Souter's kitchen in Ballarat, which looks out into a cool, enclosed courtyard garden from the kitchen counter. Inner courtyards are important sources of cool air, which can be drawn through screens to cool the house in summer.

Whether we are designing new houses or

POND WITH WATER PLANTS, FROGS, A TORTISE OR TWO.

FRONT WALL OF SHADEHOUSE REMOVED FOR DRAWING.

FIGURE 4.1 Shadehouse design backing on to the kitchen window for both coolness and interest while doing the washing up.

modifying existing ones, we can arrange it so that we walk from the kitchen to the shadehouse or greenhouse, or with a direct view from the washing-up area **(Figure 4.1)**. Put some life into these areas; perhaps a covey of little quail. The quail run about catching insects; frogs climb out of the pond, into the leaves, and even cling to the kitchen window. If you have to stand somewhere doing tedious work, at least make it interesting. Put a few little turtles—not snapping ones—into the pond. They often disappear into the mulch, eating slugs and worms. And in warmer climates you can't beat a gecko. The average gecko is designed for greenhouses and will go anywhere: upsidedown, downside up, and round about.

The shower can be part of an attached glasshouse, releasing steam, heat and water into the growing area **(Figure 4.2)**. Used bath and shower water held in a sealed earth tank or pipes below the glasshouse floor keeps earth heat high.

The path from the garden to the entry should be designed to save housework. Tracking mud or dirt into the house is usually the problem, so it is worth whatever time it takes to raise up, camber, drain, and cap (with flagstones, pebbles, concrete, or stabilised earth) the path leading from the garden to the house entry. Just before the entry itself, a special mud-grate can be installed to scrape mud off boots **(Figure 4.3)**.

Of particular interest to the cook/gardener is the layout and inclusion of a preparation and storage room just off the kitchen, called the "mudroom" **(Figure 4.4)**. This room serves as a link between garden and kitchen, and might contain:

• Food storage areas such as pantry shelves, freezer and refrigerator for home-canned products; pickle and olive crocks; wine or beer-making equipment; dried herbs, fruits, and root crop storage; and preserved meats or fish.

• Washing and preparing areas for immediate use or for preserving garden and orchard produce; a compost bucket near the sink takes

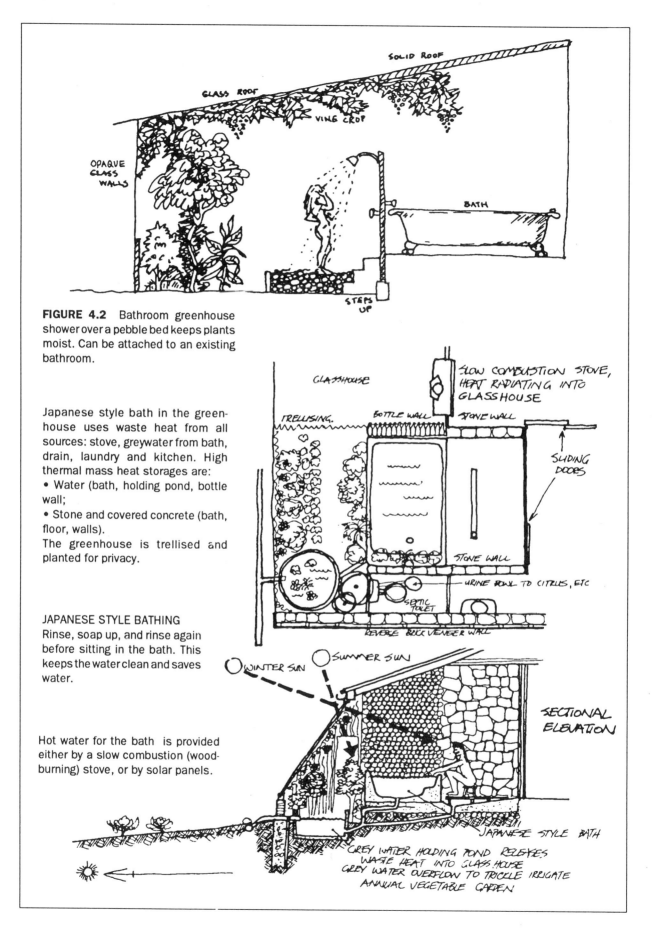

FIGURE 4.2 Bathroom greenhouse shower over a pebble bed keeps plants moist. Can be attached to an existing bathroom.

Japanese style bath in the greenhouse uses waste heat from all sources: stove, greywater from bath, drain, laundry and kitchen. High thermal mass heat storages are:
• Water (bath, holding pond, bottle wall;
• Stone and covered concrete (bath, floor, walls).
The greenhouse is trellised and planted for privacy.

JAPANESE STYLE BATHING
Rinse, soap up, and rinse again before sitting in the bath. This keeps the water clean and saves water.

Hot water for the bath is provided either by a slow combustion (wood-burning) stove, or by solar panels.

75

large leaves, roots, and vegetable tops, to be returned to garden soils.

• Dark area for growing mushrooms.

• Space for hanging wet weather gear, garden shoes or boots, and small, important food-gathering items (secateurs, knives, baskets).

• Modest home woodwork and workshop bench; tool storage.

• Cool, dry area for seed storage and desk space for garden calendars, plans, and yearly diaries.

• Firewood storage with an access flap to serve the kitchen woodstove.

FIGURE 4.3 Grate and doormat to remove mud from boots at house entry.

4.2
THE TEMPERATE HOUSE

Unless located at the sea edge (where temperatures are more even), temperate areas are cold in winter and hot in summer. Thus, house design must accommodate two different objectives. During the winter, cold must be kept outside, and heat in. During summer, heat must be excluded and the house opened up to cooling evening breezes. Energy-efficient houses can accommodate both goals through careful design. The essentials of a well-designed temperate house follow.

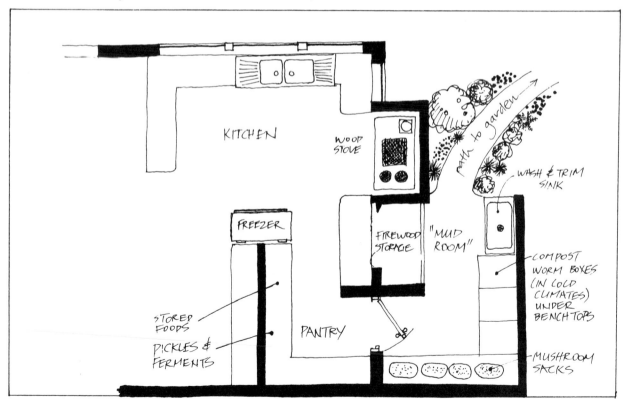

FIGURE 4.4 The mudroom as a preparation and storage area, linking the garden to kitchen.

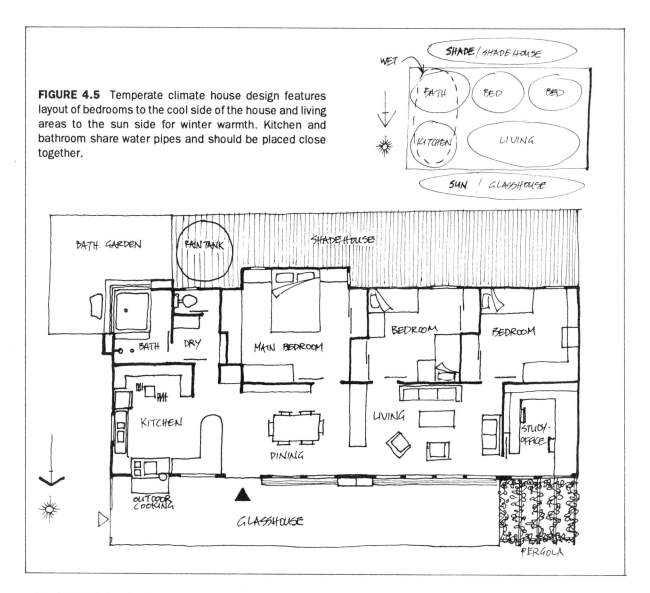

FIGURE 4.5 Temperate climate house design features layout of bedrooms to the cool side of the house and living areas to the sun side for winter warmth. Kitchen and bathroom share water pipes and should be placed close together.

■ HOUSE PROPORTIONS AND WINDOW PLACEMENT

Houses should be no more than two rooms (10 metres) deep, with east/west axis 1.5 times longer than the north/south axis. The east/west axis should face the sun (north in the southern hemisphere, south in the northern). House layout is planned so that bedrooms or other little-used rooms are placed on the shade-side of the building, while activity areas are located on the sun-facing side for winter warmth (**Figure 4.5**).

The eaves of the house, and the height or depth of windows, are designed so that winter sun strikes directly into the house through the windows (onto a slab floor or inner wall of brick or other heat-holding mass), but does not enter in summer (**Figure 4.6**).

Smaller windows are located on the east side for morning sun. There are few windows on the west and shade-facing sides of the building, as the western aspect builds up heat in summer and glare from snow in winter. Windows are fitted with heavy, floor to ceiling, pelmetted curtains, which are closed on winter evenings. In summer, windows are left open at night to allow the house to cool, then shut in the morning. Rolled bamboo blinds placed *outside* the east and west windows prevent sun strike into the house on particularly hot days.

The shade aspect (south in the southern hemisphere; north in the northern) accommodates a shadehouse with a well-insulated window opening into the house to bring cool air into the house during hot summers.

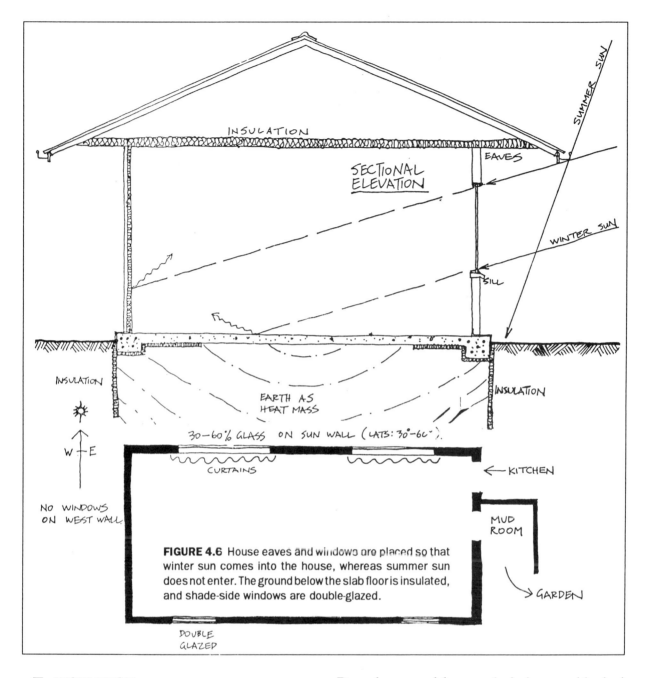

FIGURE 4.6 House eaves and windows are placed so that winter sun comes into the house, whereas summer sun does not enter. The ground below the slab floor is insulated, and shade-side windows are double-glazed.

Labels in figure: INSULATION, SECTIONAL ELEVATION, EAVES, SUMMER SUN, WINTER SUN, SILL, INSULATION, INSULATION, EARTH AS HEAT MASS, 30–60% GLASS ON SUN WALL (LATS: 30°–60°), CURTAINS, KITCHEN, NO WINDOWS ON WEST WALL, MUD ROOM, GARDEN, DOUBLE GLAZED, W E

■ INSULATION

The house is well-insulated (floors, ceiling, and at least 1 metre in the ground around the house perimeter if using a concrete slab floor). The ground insulation is of rigid foam, which is only 4-5 cm thick.

Usually, heavier or thicker insulation is placed in the ceiling to keep warm air inside during winter months.

Vents are placed in attics and crawl spaces to control damage by condensation and to allow excess heat to escape during the summer. Draughts around doors and windows are blocked with weather-stripping.

Sun coming in through the windows in winter strikes a thermal mass such as a concrete floor, brick or rock wall, or water tanks. These act as heat banks which re-radiate heat in to the house at night. During summer they remain cool during the day if exposed to cool night air (windows open at night).

Outbuildings, adjoining the house on the shade-side or wind sector, insulate the house from cold winter winds.

NATURAL INSULATION MATERIALS

There are many excellent heat insulators found in the natural world, some of which have been tried in refrigeration, house building, or noise suppression. Few are flammable, or they can be treated to smoulder rather than flame by using calcium chloride. Some are pest-immune (e.g. sawdust from trees known to be pest-immune), but all can be treated for pests using such natural products as white cedar tree leaf powder or oil, derris dust, and similar substances.

A list of potential natural insulators are as follows:

• Sawdust: was used widely in old-style refrigeration rooms and ice-houses; a vapour barrier is needed, or the sawdust bagged in plastic and sealed.

• Wool: excellent for fire retardation and warmth, as are felt and wool products or furs.

• Feathers: used for centuries in bedding and are useful in walls, ceilings; they need to be enclosed in mesh bags to keep from blowing around in draughts.

• Kapok: extensively used for bedding, also in walls and ceilings.

• Seagrass (*Zostera, Posidonia, Ruppia*): Dried and partly-compacted; a traditional wall and roof insulation of low fire risk.

• Straw: a good insulator where fire is not a problem; now available commercially as fireproofed compressed board sheets for ceiling (wire-bound or stitched).

• Cork: as granules, slabs, tiles, pressed blocks.

• Fibrous waste: e.g. from processed licorice root and the fibre of coconut husk (coir), which also suits roll matting; coir is pest-immune in most cases.

• Paper: shredded waste paper soaked in 1 part borax and 10 parts water is a good insulation.

• Balsa: both the wood itself and the cotton from the seed pods have been long used as insulation. As the tree grows fast in the humid tropics, it is a good use of land to produce insulation blocks.

Insulation is essential in temperate to cold areas; however, care must be taken to maintain adequate ventilation, especially in cases where houses are sited near areas subject to radon emissions (a gas emitted from granite, dolerite, and most igneous rocks).

■ PLANTING AROUND THE HOUSE

Deciduous trees planted on the sun-side and the east side of the house allow winter sun to penetrate in autumn/winter. In full leaf, they shade the house in summer, preventing sun strike on all parts of the roof. Trellis of deciduous vines (wisteria, grape) located at strategic places around the house provide some shade effect while large trees are growing (**Figure 4.7**).

The west and shade-facing walls are available for evergreen trellis and shrubbery to protect these areas from exposure (heat in summer and cold winds in winter).

The goal of house design is to reduce or eliminate the need for electric or gas energy input for inside heating and cooling. Because the sun heat is regulated and stored in the heat masses of floors, walls, and water tanks, and draughts are excluded, then the slight heat yield from body warmth, cooking, and a small wood-burning stove is all that is needed to keep the air space warm.

In areas of severe cold winters, specific house problems are heat costs, snow load, condensation, cold winds, and damp. The types of houses found in these areas are conjoined, multi-storey, steep-roofed, radiant-heated, and insulated. In rural areas, houses are attached to barns, and if possible, with earth insulation up to 1.2 metres. Basements or cellars are common for coal/wood storage, worm beds, large manure pit (below barn) and root storage.

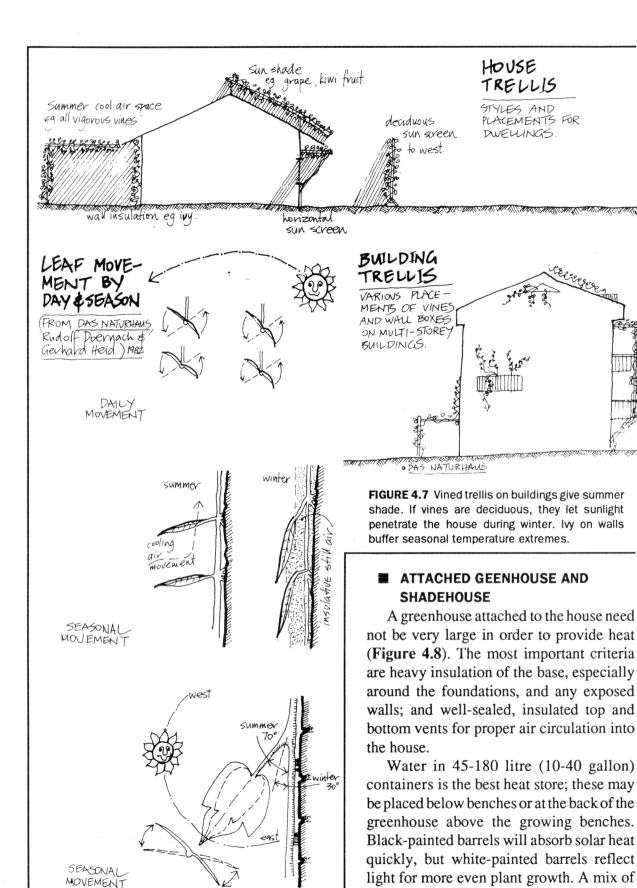

HOUSE TRELLIS

STYLES AND PLACEMENTS FOR DWELLINGS.

Summer cool air space eg all vigorous vines

Sun shade eg. grape, kiwi fruit

deciduous sun screen to west.

wall insulation eg ivy.

horizontal sun screen

LEAF MOVEMENT BY DAY & SEASON

(FROM DAS NATURHAUS, Rudolf Doernach & Gerhard Heid.) 1982

DAILY MOVEMENT

summer

winter

cooling air movement

insulative still air

SEASONAL MOVEMENT

west

summer 70°

winter 30°

east

SEASONAL MOVEMENT

○ BY COURTESY DAS NATURHANS

BUILDING TRELLIS

VARIOUS PLACEMENTS OF VINES AND WALL BOXES ON MULTI-STOREY BUILDINGS.

○ DAS NATURHAUS

FIGURE 4.7 Vined trellis on buildings give summer shade. If vines are deciduous, they let sunlight penetrate the house during winter. Ivy on walls buffer seasonal temperature extremes.

■ ATTACHED GEENHOUSE AND SHADEHOUSE

A greenhouse attached to the house need not be very large in order to provide heat (**Figure 4.8**). The most important criteria are heavy insulation of the base, especially around the foundations, and any exposed walls; and well-sealed, insulated top and bottom vents for proper air circulation into the house.

Water in 45-180 litre (10-40 gallon) containers is the best heat store; these may be placed below benches or at the back of the greenhouse above the growing benches. Black-painted barrels will absorb solar heat quickly, but white-painted barrels reflect light for more even plant growth. A mix of the two might be best.

Double-glazed panels are the most du-

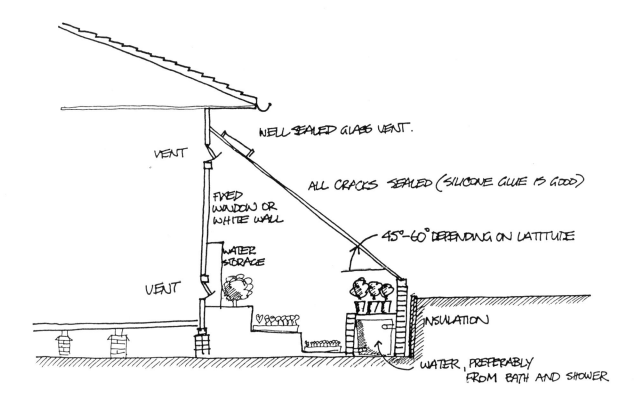

FIGURE 4.8 A greenhouse on the sunside of the house will help with house heating, especially in cold winter climates. Vents are essetential for bringing heat into the house during winter and cooling it in summer.

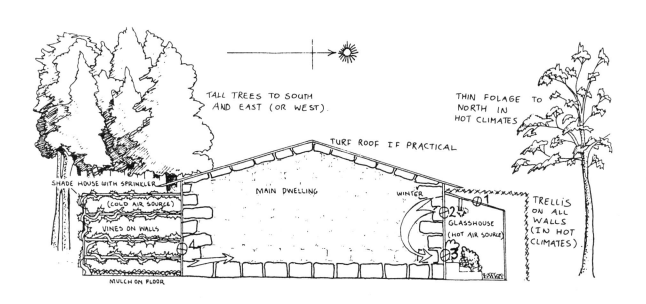

FIGURE 4.9 Cool air circulates from the shadehouse into the house, drawn in by the greenhouse. Deciduous vines (grape) are on the sun side, whereas evergreen vines are on the shade side.

rable and efficient, holding heat in longer than single sheets of glass. Wooden frames are used to prevent heat escape (metal frames lose heat too quickly).

To circulate a cooling breeze in the summer (usually in the evening), an attached shadehouse on the shade-side of the house is an important part of the greenhouse system. **Figure 4.9** shows how this system works. In summer, when the house is too hot, open Vent 1 at the top of the greenhouse; air escapes, drawing in cool air from Vent 4, over the damp mulch and through the vine-covered and ferny shadehouse, where a fine spray or drip of water on the mulch keeps the air cool. In winter, close Vents 1 and 4, open

Vents 2 and 3, so that by day warm air from the greenhouse circulates in the insulated rooms. Close at evening, trapping warm air.

Water tanks can be vine covered in the shadehouse as a cool air/water block. Both the shadehouse and greenhouse yield food for the family while cutting down on fuel costs.

■ HOUSE MODIFICATION

Many already-built houses must be modified so that they are as energy-efficient as possible. The main problem lies in the often perverse arrangement of older houses, which face the road rather than the sun, and in the mania for glass windows on all outside walls. We can summarise ways to make older houses more

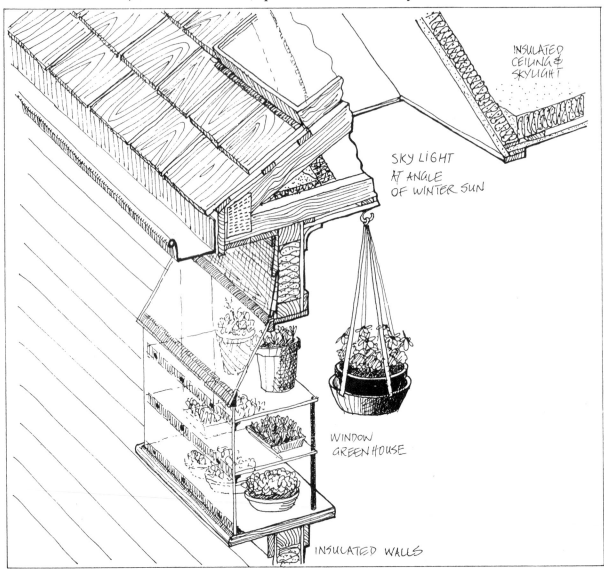

FIGURE 4.10 Attached windowbox greenhouse and skylight. Note insulation to retain heat and deter cold.

energy-efficient, in order of most concern:

• Careful draught-proofing of all doors and windows. Sealing all cracks is essential to prevent heat leaking from the house or cold air entering.

• Insulation of walls and ceilings; this alone will reduce heating and cooling bills by 50%.

• Attaching a greenhouse to the sun-side, if possible; even a window greenhouse and skylight is an improvement as it brings in sunlight and green growth (**Figure 4.10**). Double-glass is essential in temperate areas, and in cold regions the greenhouse needs to be closed off from the rest of the house.

• Adding heat mass as concrete slabs, tanks, and brick or stonework within the greenhouse or insulated warm rooms.

• Attaching a shadehouse to the shade-side in hot summer climates to draw cool air into the house, saving on air conditioning.

• Placing a solar hot water heater on the roof to reduce or eliminate fuel-powered water heaters.

• Using vegetation for microclimatic control, e.g. planting trees in a sun trap shape, attaching trellis or shrubberies to the shade and western aspects, planting deciduous trees or vines on the sun-side, and placing windbreak trees in the wind sector.

Well-designed homes are cheaper to maintain than houses which need expensive energy-consuming heaters and air conditioners, and enable people to survive in warmth and comfort without recourse to oil-based fuels. It is no longer necessary, or even sensible, to build any type of house other than one which saves or generates energy.

Subtropical and cold/arid house design is similar to temperate design, as temperatures can get down to freezing in almost all areas except mid-slopes and above. However, the subtropical house can also have some of the features of the tropical house.

The humid tropics are usually more subject to periodic catastrophe than the temperate lands (with the exception of fire); thus the only safe long-term house sites are:

• Above the reach of tsunami (tidal wave).

• Sheltered from cyclone and hurricane tracks.

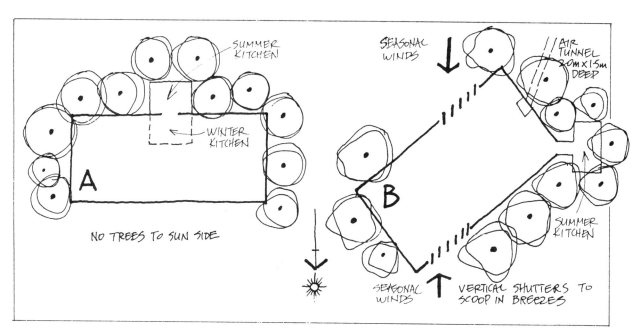

FIGURE 4.11 Siting houses in (**A**) the subtropics, where orientation is towards the sun because of cool winters, and (**B**) the tropics, where the orientation is towards cooling breezes and all-around shading.

• Above valley floors subject to mud-flow or volcanic ash flow.

• On ridge points or plateaus out of the path of rock or mud slides triggered by clear felling, torrential rain, or earthquake.

• Inland from easily-eroded sandy beaches.

The main aim in hot, humid regions is to prevent the sun from striking the house, and to dissipate built-up heat (from humans, appliances, cooking) from the house. Thus, shading the house and orienting it to catch cooling breezes are primary considerations (**Figure 4.11**). Find sites where moderate winds blow, where forests or deep valleys help shade and cool the house, or, in strong wind areas, where the structure is protected from severe winds by forest, earth ridges, or is naturally sheltered in narrow cross-wind valleys.

The shape of the house is elongated or irregular to increase surface area. There are no solid, insulated walls to accumulate heat, and houses are most often open-plan style for air circulation. If internal walls are used, they are made of light materials (matting, louvres, net-

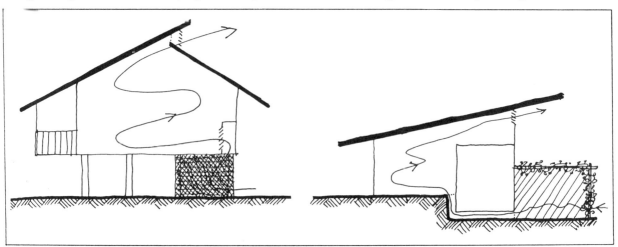

FIGURE 4.12 Vented ceilings allow hot room air to escape, and cool trellis air to enter.

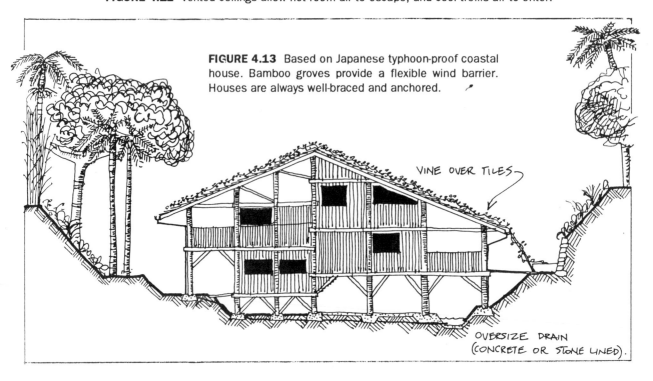

FIGURE 4.13 Based on Japanese typhoon-proof coastal house. Bamboo groves provide a flexible wind barrier. Houses are always well-braced and anchored.

VINE OVER TILES

OVERSIZE DRAIN (CONCRETE OR STONE LINED).

84

ting) and stop short of the ceiling to allow free air flow.

Ventilation is essential, through the placement of windows (with vertical louvres acting as air scoops) and roof vents. Or a shadehouse can be added to the shade side of a house and cross-ventilated to a well-vented ceiling or solar chimney (**Figure 4.12**).

There are wide verandahs on all sides of the house, often supporting vine crop. (In the subtropics, the verandah is partially omitted on the sun-side of the house in order to let winter sun penetrate into the house.

Vegetation shades the house; particularly useful are tall trees with smooth trunks (no dense branching) such as palm trees that grow up past the verandah and overshade the roof. Care must be taken, however, not to completely surround the house with plants, as dense vegetation blocks cooling breezes and raises the humidity around the house. Grass rather than paved areas prevent heat reflection to the walls or eaves.

Heat sources such as stoves and hot water systems are detached from the main structure; many traditional houses in the tropics have outdoor kitchens for summertime cooking.

Insect screens are on all doors and windows in areas with high concentrations of mosquitoes and other noxious insects.

The roof is painted white or is reflective, sending heat back into the atmosphere. Roof angles are steep both to shed heavy rain and to withstand strong winds. In known hurricane areas, very strong cross-bracing, deep ground anchors, and strapped timbers are necessary. Large bamboo groves placed to windward will bend to the wind without breaking, protecting the house (**Figure 4.13**).

A hurricane cellar or stone-concrete core (e.g. bathroom area) can be built inside or outside for emergencies, and should have a concrete roof. Alternatively, an earth cave or trench, preferably roofed solidly, can be made outdoors. All windows and doors are provided with shutters and solid wooden locks (drop-in bars).

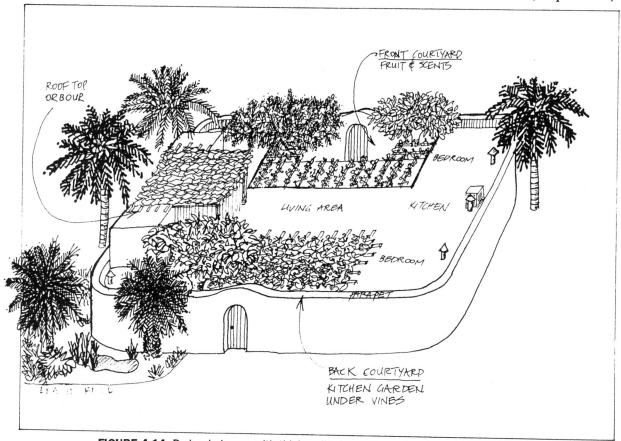

FIGURE 4.14 Drylands house with thick walls, internal courtyards and overhead trellis.

There are several designs for the dryland house, depending on seasonal temperatures. Some dryland areas experience cold winters and hot summers, while others (closer to the equator) enjoy mild winters.

In shape and orientation, the basic temperate-area house design applies for the hot, arid regions with cold winters. However, there is more emphasis on providing cool air sources:

• **Internal courtyards**: preferably latticed or shaded overhead by trees (**Figure 4.14**). They are even more effective if they are two or more storeys high and naturally shaded by the building, although small courtyards with shadecloth can also be added to single-storey houses.

• **Extensive fully-enclosed vine arbours** with mulched floors and trickle-irrigated (**Figure 3.7**) These suit single-storey dwellings. Arbors need to be about 30% of total floor area to provide cool air; hanging house plants aid in cooling, as does a water tank.

• **Earth tunnel**: a 20 metre long, 1 metre deep trench sloping downhill to the house. In the tunnel, large unglazed pots full of water, pans of wet coke, or curtains of coarse fibreglass weave can be drip-fed to provide evaporative cooling. Cool humid air continually falls through these tunnels to the house rooms (**Figure 4.15**).

• **Induced cross-ventilation**: This is most easily achieved by fitting a black-painted sheet-metal solar chimney to open from ceilings or roof ridges. As these heat up, they effectively draw air into the rooms from any of the above cool-air sources, and create a cool air flow in living areas (**Figure 4.12**).

For both heat and cold control, thick walls, edge-insulated floors, draught-proofing of doors and windows, insulated ceilings, and efficient cross-ventilation are all important ways to moderate the extremes of daily and seasonal temperature typical of many desert areas. White-painted exterior walls help to reflect excessive heat, and well-placed shade trees, palms, vine trellis, and courtyard ponds or fountains assist in buffering heat extremes.

As in tropical climates, an energy-saving design feature is to locate an outdoor screened-in summer kitchen part-roofed under a thickly trellised area, where occupants can spend most of the day out of doors.

In many dryland areas, rooftops are flat and contain many of the features usually found around the house in temperate or tropical areas. These include header tanks for 1-2 weeks' water supply; laundry and clothesline; pigeon pens for eggs, squabs, manure; grain and vegetable drying flats; evening sitting areas; and potted plants (**Figure 4.16**).

It is particularly important in desert areas to

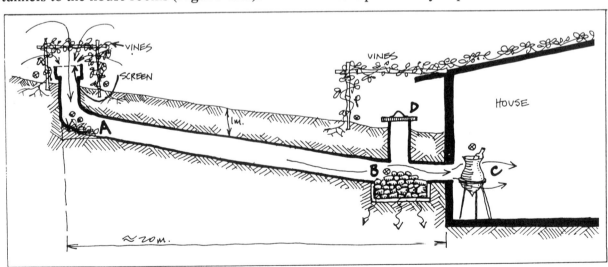

FIGURE 4.15 Earth tunnel provides cool humid air to dry desert houses. Tunnel slopes down to the house, has a shaded intake, moist cinder bed, and unglazed pot of water at outlet; length 20 metres.

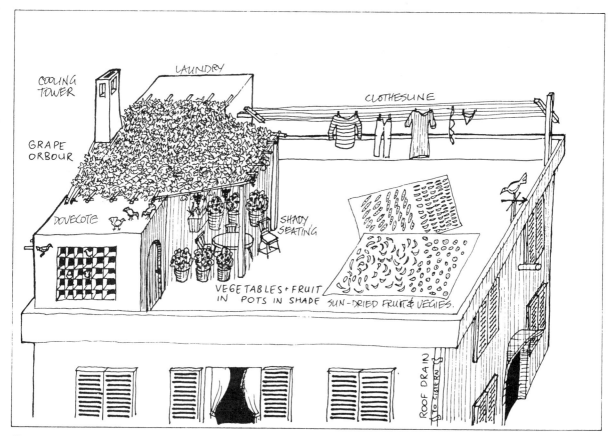

FIGURE 4.16 Rooftop "furniture" of hot-dry climates, where houses are often joined and there is little or no yard available. Many yard functions therefore take place on the roof.

conserve household water. Modest water use is easily achieved if efficient shower heads are used for washing, and both shower and handbasin or laundry water is first diverted to the flush tank of toilets (if sewage lines are provided), or to the garden. To get shower water to a flush toilet cistern, the shower and handbasin can be raised a few steps above floor level, and a low-level cistern used (**Figure 4.17**). All roof areas should collect water into storage tanks, which are located on the shade-side of the house, under trellis, to provide cool water for drinking.

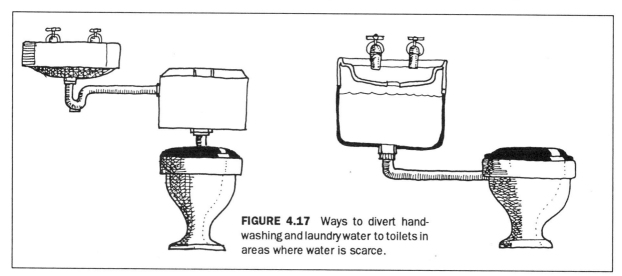

FIGURE 4.17 Ways to divert hand-washing and laundry water to toilets in areas where water is scarce.

87

■ UNDERGROUND HOUSING

In both ancient and modern times, caves and underground houses were the preferred dwellings in deserts (particularly those with mild winters). Their practicality depends on the location having softish rock, or a softer strata below a calcrete or ferricrete "ceiling". Cave houses can be totally below ground, with skylights, but they are more commonly built with one wall facing out from the open (sunny) side of a hill. Sunrooms can be built out in front of the underground rooms, or front rooms built on as a facade.

Decorative facades may be built at the entry, and shaded by grape trellis. Where occasional rains are expected, sections of the hill slope above the cave can be sealed with concrete as a roof or water runoff area for water cisterns; this

FIGURE 4.18 Earth bermed house for arid climates keeps house insulated and cool. Vines can shade sunward walls.

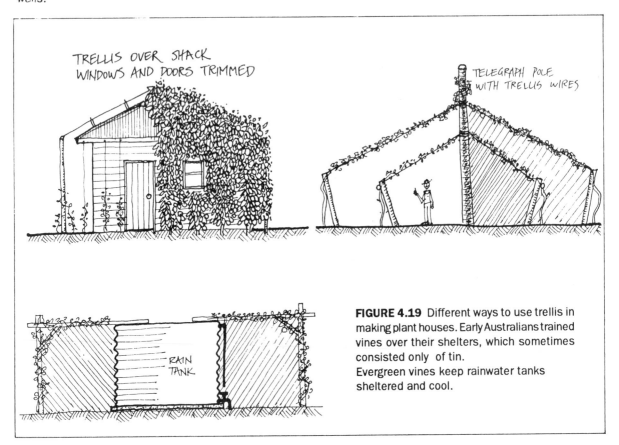

FIGURE 4.19 Different ways to use trellis in making plant houses. Early Australians trained vines over their shelters, which sometimes consisted only of tin.
Evergreen vines keep rainwater tanks sheltered and cool.

also strengthens the strata above the rooms and prevents water seepage into the cave.

A cool house for deserts, duplicating cave conditions, is the dwelling with earthbanks built up to the eaves (and if required, over the roof) as seen in **Figure 4.18**.

The cool conditions of caves, brick tanks, fire refuges and root cellars offer great advantage in storing and preserving a great variety of goods. Cool caves greatly prolong the life of citrus, root crop and leaf crop in storage, and are cool air sources in summer.

Also, a cave near the house has value as a family refuge in catastrophic wind, fire, war, or heat wave. Such structures may be dug into banks. Also possible are underfloor cellars entered from floor traps or outside cellar doors, or above-ground structures of ribbed steel or pipes earthed over for protection. Radiation from fire is prevented by a "T" shape or a "dogleg" in the entry of shelters.

4.5
PLANT HOUSES

There are varying degrees of integration of house and plants: from the totally grown house to vine-covered or sod-roofed conventional structures.

Rudolf Doernach, in Germany, has designed a house with a light steel and timber frame. This frame is grown over with evergreen, waxy-leaved climbing plants (several species of ivy, geranium, and coastal climbers suit this description). Only doors and windows need to be kept clear of vine, and as the structure is designed to take creepers, trimming is unnecessary. The building is igloo-like in form, a necessity for cold winters.

In the early part of the century, settlers in the arid area of Western Australia built an outsize structure over their sheet iron buildings. On this they trained climbing plants so that eventually the entire building was covered (**Figure 4.19**) to moderate the extremes of hot and cold.

This technique can also be used in any climatic zone, with appropriate climbing species. In mild to warm temperate areas, examples of vines are as follows:

Fast-growing deciduous vines: kiwifruit, Chinese trumpet creeper, sweet woodbine, Chile jasmine, Virginia creeper, grape, wisteria.

Edible fruit climbers: kiwifruit, passionfruit (banana passionfruit withstands light frosts), grape.

Self-clinging climbers for brick and stone: Cross vine, cat's claw creeper, climbing fig, English or variegated ivy, Mexican blood trumpet.

■ SOD ROOFS

Sod roofs are another plant/house system, and may be newly constructed, or rolled over strong existing structures, using a plastic film stapled below as a moisture barrier. The metal cog carries water to the spout, while leaves drop off (**Figure 4.21**). The slotted angle or log (indispensable on steep roofs) holds the sod from slipping.

Trials of smaller roofs on sheds and animal houses are probably the best way to get the technique and species right, and as the weight of winter sod roof is great, loads must be carefully calculated.

I can always bring a nervous titter from an Australian audience by suggesting that they shift their lawn onto their roof. But I am being fairly serious, as sod roofs are great active insulators, and any strong (or strengthened) roof would take sod, either as ready-rolled lawn in humid areas, succulents such as iceplants or pigface in dry areas, and with daisies, bulbs, and herbs elsewhere.

Evapo-transpiration, plus judicious watering, keeps the summer heat out. In winter the air and foliage keeps winter cold at bay. Sod roofs act, in fact, like ivy on walls. Neither increase fire risk to the house.

For weak existing roofs, especially those of zinc or aluminium cladding sheet, ivy or light vines over the roof serves as a light insulation, provided the guttering is adapted as shown for sod roofs.

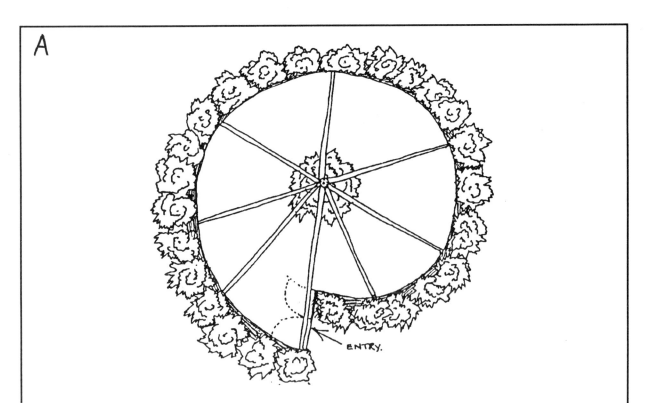

FIGURE 4.20 Cheap livestock shelter: plan (A) above and elevation (B) below. Concrete or tiled slab on ground, with a spiral of bamboo or poplar centred on an existing tree or pole. Walls are woven and ivy-covered to the roof.

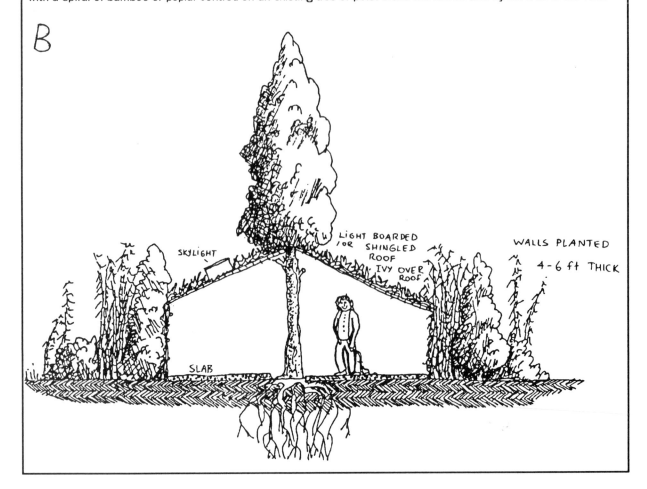

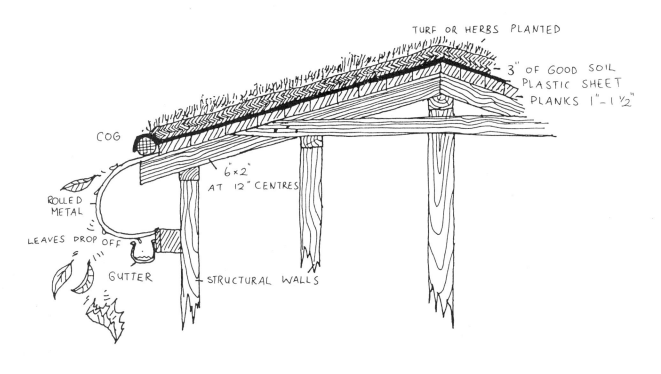

FIGURE 4.21 Sod or turf roof construction.

WASTE RESOURCES FROM THE HOUSE

The "waste products" of a house are all too often viewed as disposal problems rather than as resources. These waste resources are wastewater from showers, sinks, and laundry; sewage; food scraps; and paper, glass, metal and plastic garbage.

Glass and metal can be recycled, while plastics are kept to a minimum if you take your own shopping bags to market. Newspapers and office papers are used as a mulch barrier in gardens and orchards, or soaked and fed to worms (in limited quantities).

The most important products are wastewater and sewage, and these are treated in different ways according to climate and preference. In drylands or dry seasons, where water is at a premium, sink and shower water is diverted to a grease-trap and from there used to irrigate garden beds. Handbasin water can also be used to fill the cisterns of flush toilets, thus performing a double-duty. All roof water is carefully diverted to storage tanks.

In the tropics, where summer downpours are frequent and storage tanks are easily filled, roof water should be directed away from the house and garden into gravel-filled channels and planted swales to prevent erosion of the driveway, garden, and house surrounds. During the dry season, when rain is infrequent, roof gutters lead to storage tanks for drinking water.

Sewage from flush toilets can be routed via a septic tank or methane generator to plant systems (orchard crops) as shown in **Figure 4.22**. Compost from dry toilets is buried beneath trees; or, in the case of moveable pit toilets, a tree is planted on top of a closed pit.

Food scraps are fed to animals (including worms) and their manures used in the garden. Alternatively, scraps are composted or even directly buried into garden beds, although these will become hot under the ground as they break down, so caution must be taken not to plant immediately in the area. Thus household waste products are used in the system to produce food and nutrients to plants and animals.

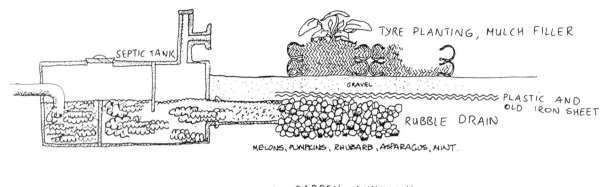

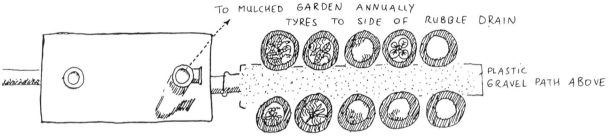

FIGURE 4.22 Disposal of septic tank effluent via tyre-planted rows.

4.7
TECHNOLOGICAL STRATEGIES

Modern western houses use about 5 kilowatts of power, but by using a combination of strategies, especially good house design, solar hot water, insulation, and sensible, responsible behaviour, this can be reduced to 1 kilowatt or less, permitting much smaller energy systems to be installed for peak loads. The general categories for technological energy conservation in the home may be summarised as follows:

Climate control: space heating and cooling
• Woodburning stoves: fast-burning, massive radiant heaters, or slow-burning, efficient cast-iron stoves.
 • Greenhouse attachment for winter heating.
 • Shadehouse attachment for summer cooling.
 • Trellis systems for sun deflection; cooling.
 • Conducted heat: usually large under-floor systems using water pipes or electrical wires connected to waste heat.

Cooking and Cookstoves
• Wood-fuelled cookstoves (best in cold temperate climates) provide heat as they cook.
 • Gas/propane stoves suit hot and hot-humid climates; a gas system leaves open the potential to use methane from biogas digesters using sewage and other wastes.
 • Solar cooking units are divided into two types: reflective parabolic arcs focusing onto one point and solar ovens (home-made) which are glass-fronted insulated boxes lined with reflective aluminium foil. Both types must be moved by hand to follow the sun unless fitted with a solar tracking device.
 • Insulated container cooking is an effective method for items which need a long cooking time. Essentially, one boils a pot (of stew, casserole, beans, soup) for between 1-3 minutes. The hot pot and its contents are then immediately transferred to an insulated box where it continues to cook (**Figure 4.23**).

Hot Water Supplies
• Wood cooking or heating stoves with a 18cm copper or stainless steel tube loop in the fire-box (to the back or one side) will provide hot water to an insulated storage tank.
 • Solar collectors on the roof can be pur-

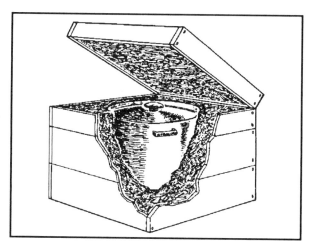

FIGURE 4.23 Insulated box cooker for slow-cooking foods (legumes and grains).

chased commercially or hand made, and include flat plate, bread-box, and cylindrical collectors.

Electricity and Lighting
• Solar photovoltaic cells and storage batteries are used to power house lights and appliances.

• Wind power or small-scale hydro-electric in appropriate locations provide for all lighting and appliance needs.

• Energy-conserving and long-lasting lights such as low pressure sodium lamps are recommended for rooms that are in almost constant use (kitchens).

• Gas and kerosene (mantle and wick lamps) lighting is useful for those in the country who do not need much light or haven't the funds to purchase more expensive systems.

Washing and Drying Clothes
• In Australia and Europe small hand-operated pressure washers (Jordashe, Bamix, Presawash) are run by water pressure through a hose; these have a small capacity and suit individuals or couples.

• For bigger families and communities, a shared coin-operated washing machine saves money.

• Clothes can be dried on a clothesline, in a greenhouse or similar airy and roofed area, or, for small items, in an insulated cupboard sur-

rounding an uninsulated hot water cylinder. In wet temperate regions a rack above the wood-stove is traditionally used for clothes drying or, in autumn, for drying herbs, flowers, or seed-heads (**Figure 4.24**).

Refrigeration and Cooling, Food Drying
• Gas and kerosene refrigerators are available, and are usually small and efficient. A large photovoltaic system, wind power, or hydro-electricity easily powers a refrigerator.

• An airy, screened cupboard, open on one side to the shadehouse in temperate areas, can be used to store fruits and vegetables, eggs, and anything else that does not require cold refrigeration.

• For drying fruits and vegetables, a solar food dryer or a semi-empty greenhouse in summer will do the job.

Water Conservation
• Water tank off barn/garage roof is ideally located uphill from the house for gravity flow.

• Hand-basin water is used to flush toilets; or hand-basin and shower wastewater diverted to garden/greenhouse.

• Low water use shower nozzles are commercially-available.

• Toilets with two flush modes (11 litres for solids, 5.5 litres for liquids) are now used in most new homes in Australia.

• Compost toilets or pit privies use no water, and provide composted manures for use around trees and shrubs.

Enormous savings of national and international petrol, coal and gas are achievable if houses and communities are designed and equipped for energy conservation. Most of the home energy systems above are non-polluting and beneficial. Given the atomic fallout and acid rain from reactors, power stations, and cars, our only possible future is to develop clean energy and to reduce energy use; that is, the ultimate saving we may make is to our own lives, and those of the forests and lakes of the planet.

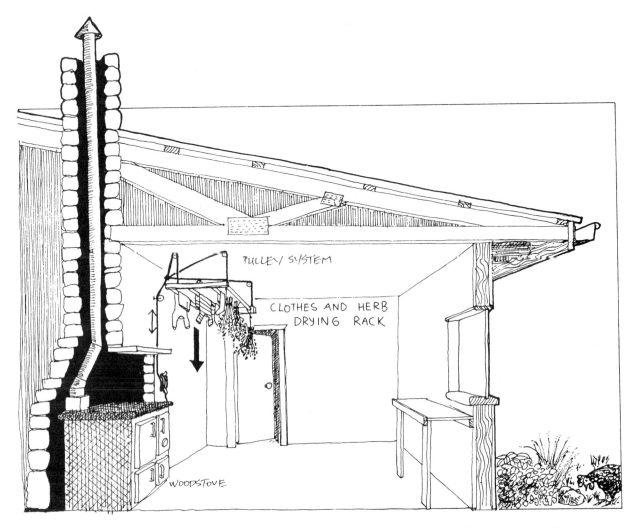

FIGURE 4.24 Pulley system for drying rack above the woodstove dries clothes in cold, wet weather

4.8
REFERENCES AND FURTHER READING

Corbett, Michael, and Judy Corbett, *A Better Place to Live*, Rodale Press, 1981.

Farallones Institute, *The Integral Urban House*, Sierra Club Books, San Francisco, 1979.

Leckie, Jim, *et.al.*, *More Other Homes and Garbage: designs for self-sufficient living*, Sierra Club Books, 1981.

Technical Assistance Group, *Low Cost Country Home Building*, Dept. of Architecture, Univ. of Sydney, Hale & Iremonger, 1983.

Vale, Brenda and Robert, *The Autonomous House: design and planning for self-sufficiency*, Thames & Hudson, 1975.

Vasella, Alessandro, *Permaculture or the End of the Myth of the Plough*, Pamphlet

CHAPTER 5

Home Garden Design

5.1 INTRODUCTION

Zone I is that area closest to the house, starting just outside the kitchen door, and contains the annual garden, small important perennials, espalier or miniature fruit trees, seedling and nursery beds, and small quiet animals such as rabbits and pigeons. It is the zone which we visit daily, and which is intensively planted and controlled.

The size and shape of Zone I depends mainly on site acreage, access, schedules, and time available, so that if there are daily visits to the barn or laying shed to collect eggs, then Zone I would stretch from the house to the barn. Those with time to devote to the land and a large family might have a large Zone I, while those who work offsite might limit their Zone I to a 4-8 square metre section just outside the door.

The structures associated with Zone I are the greenhouse and shadehouse (discussed in Chapter 3), potting shed, propagation frames, composting area, clothesline, barbecue pit, and garden storage area. Other structures might include a pigeon loft on the roof or off the house to collect manure and raise squabs; small pens for rabbits or guinea pigs; and a workshop.

When planning Zone I, we need to look at:

• **Climate and aspect**: From which direction does the wind blow? Which is the sun side? Shady areas? Where do frosts strike?

• **Structures**: Where can structures be placed so that they fulfill two to three functions? Can they be used as: water collectors; trellis supports; windbreaks; food production areas?

• **Access**: How should access be arranged: roads, entries, clothesline, play area, woodpile, barbecue, pathways, mulch heaps?

• **Water source**: What are the garden water sources: tanks, hoses, greywater from the house, and how is water to be distributed (sprinklers, drip irrigation)?

• **Animals**: What small, useful animals should be in Zone I, and what systems need to be provided for them (food, shelter, water)? How can large animals be excluded by hedges or fences?

Everything should be considered in relation to each other, so that the products of one element provide for the needs of another.

If you are having trouble knowing where to start, always start at the doorstep, as the house provides a central focus and an edge from which to work outwards. If you need to, first make a layout map of the house, trees, fences, pathways, and any other existing structures or features. Then decide what you want close to the house (garden structures, garden beds, small animals, ponds, etc.) and place them according to the basic energy-conserving rules.

5.2 GARDEN LAYOUT

The garden is fully-mulched, with its soils aerated and humus-rich. Plants are constantly being recycled; tops are eaten, leaves discarded; green manures are turned into the soil to provide nutrients for a summer crop; some dill, carrots, and fennel are allowed to go to flower to attract parasitic wasps; and volunteer tomatoes and cucumbers from the compost heap are planted out along the fence.

There is no attempt to form the garden into

strict neat rows; it is a riot of shrubs, vines, garden beds, flowers, herbs, a few small trees (lemon, mandarin), and even a small pond. Paths are sinuous, and garden beds might be round, key-holed, raised, spiraled, or sunken.

It does not matter what methods you use to make your garden, whether you choose to double-dig your beds, or simply sheet-mulch with newspapers and straw. It's a matter of what suits you. I'm lazy—full mulch suits me. You are vigorous—double-digging suits you. Double-digging suits you now because you may be young. Full mulching, you will grow into! Technique is not a fixed thing (nor is permaculture generally); it is something appropriate to occasion, age, inclination, and conviction.

So the important thing is to lay out the garden based on frequency of visits and size of crop, and to allow a range of plants for greater insect control. Even when designing a small area such as a garden, we can follow the general permaculture principle of placing planting beds depending on how many times the beds are visited.

■ KITCHEN DOOR CULINARY HERBS

Imagine a clump of parsley 6 metres away in the main garden. You've just made soup and want to season it before serving. It's raining outside, and you are in your furry slippers. There is *no way* you are going to rush out and get that parsley! It and many other herbs in the garden remain unharvested because they are too far away. But if we have a herb bed just outside the kitchen door, harvesting fresh herbs is no problem.

A herb spiral (**Figure 5.1**) accommodates all the basic culinary herbs on a mound of earth on a 1.6 metre wide base, rising to 1 or 1.3 metres high. This spiral gives variable aspects and drainage, with sunny, dry sites for oil-rich herbs such as thyme, sage, and rosemary, and moist or shaded sites for green foliage herbs such as mint, parsley, chives, and coriander. At the bottom is a small plastic-lined pond in which watercress or water chestnut can be grown. The herb spiral is conveniently watered by one sprinkler placed at the top.

■ SALAD CLIPPING BEDS

These beds, not far away from the herb spiral, are narrow and close to the house. In them go more herbs (those that don't fit onto the herb spiral, or which you want to grow in quantity) and small salad herbs and greens such as garden cress, garlic chives, shallots, and mustard greens, which can be cut with scissors. They are very fast-growing throughout spring and summer, and yield a large quantity of greens. They are often visited, watered, cropped, and mulched to restore surface humus (**Figure 5.2a**).

■ PATHSIDE PLUCKING VEGETABLES

These are the useful, long-bearing vegetables for salads or cooking that we can either cut, or pull leaves from for months of yield. Most are transplanted from a seedling bed, and comprise

FIGURE 5.1 Garden herb spiral with small watercress pond at the foot. One sprinkler waters it all.

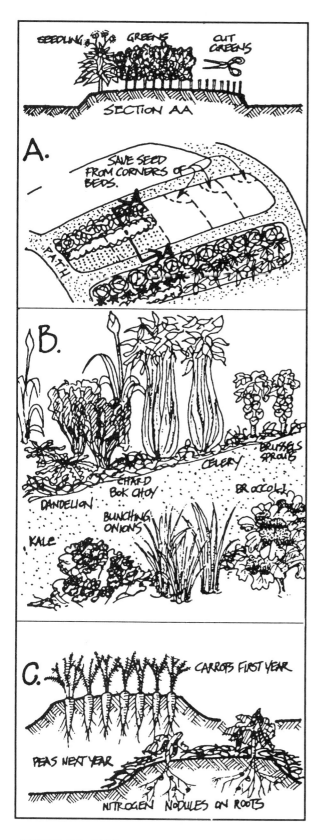

FIGURE 5.2 GARDEN BEDS
(A): Narrow beds for clipped greens **(B)**: Pathside vegetables (fill any holes with garlic, chives, parsley, etc.) **(C)**: Rotating crops each season or year.

such vegetables as Brussels sprouts, silver beet (Swiss chard), celery, bunching onions, broccoli, kale, mustard, spinach, and Florence fennel. Capsicum and zucchini are also vegetables which can be picked frequently.

These vegetables are along the pathside, and are constantly being removed, transplanted, and replanted. Most often, a leaf or stalk is taken for salads or stir-fried vegetable dishes; it is rare that the whole plant is harvested. Some are left to seed themselves in the garden (**Figure 5.2b**).

■ NARROW BED PLANTS

Now we get to the garden beds themselves, which can be separated into narrow bed plants and wide bed plants. Both types contain plants which need a long period of picking (over the summer and autumn, usually). The narrow plant beds contain plants that need more access and fairly frequent picking, and might be made up of beans, tomatoes, zucchini, carrots, peas, eggplant, salsify, broad beans, and such herbs as caraway, chervil, cumin, and chamomile (**Figure 5.2c**).

Tomato plants need a narrow bed so that they are reached and picked easily as tomatoes ripen. As they dislike wind, they can be planted in a "keyhole" bed with a sunroot (Jerusalem artichoke) surround (**Figure 5.3**).

■ BROAD BEDS

Here we plant those crops which require a long time to mature, or are harvested all at once for storage or processing. These include corn (both sweet and maize varieties), melons, pumpkins, onions, potatoes, leeks, beets, turnips and swedes. They are close-spaced, self-mulched, have no paths between them, and are block-planted. Some of these beds can also go into Zone II for main cropping.

■ BARRIER HEDGES

Around the garden, and perhaps breaking it up into manageable sections, are hedge crops. Hedges are often used as wind, weed, and animal barrier plants, and if care is used in selecting species, they can also be used as mulch sources,

97

PLANT FAVA
BEANS AS
WINTER CROP,
OR GREEN
MANURE TO
FOLLOW TOMATOES.

INTERPLANT:
A FEW MARIGOLDS,
SOME SWEET
BASIL (TO COOK
WITH TOMATOES),
AND DWARF
NASTURTIUMS.

CHIVES

BASIL

JERUSALEM ARTICHOKE WINDBREAK

FIGURE 5.3 Raised keyhole bed, densely planted, with a hardy sunflower windbreak. Such beds suit tomatoes if staked or trellised.

animal forages, nitrogen fixers, and edible crop.

Whether from neighbours' untended fences, or from the uncontrolled edge of your own cultivation, the mulched area of Zone I is under constant attack from ground invaders. Kikuyu, couch or twitch grasses reach out to smother the pampered annuals. Unless you can afford deep concrete sills under the fence, you must look to nature for the solutions.

After sheet mulching the garden (discussed later in this chapter), plant a living barrier around your protected area, and mulch it well with cardboard and sawdust or straw (**Figure 5.4**). Use vigorous, shady, or mat-rooted useful plants immune to the re-invading grasses (non-runner bamboo, comfrey); an inspection of your local area will reveal more species that do not permit the invaders' approach.

Sunroot (*Helianthus tuberosus*) planted in a band about 1.2 metres wide, acts almost immediately as a windbreak to supplement slower-growing hedges. The Siberian pea shrub (*Caragana aborescens*) fixes nitrogen, forms a thick hedge, can be grown in cold climates, and its seeds used to feed poultry. Taupata (*Coprosma repens*), planted closely together and clipped

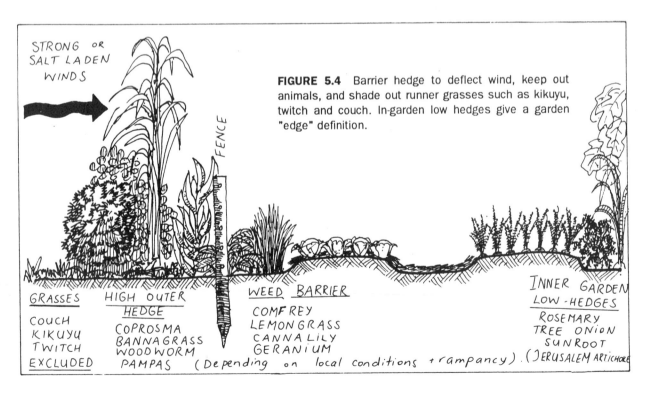

STRONG OR
SALT LADEN
WINDS

FENCE

FIGURE 5.4 Barrier hedge to deflect wind, keep out animals, and shade out runner grasses such as kikuyu, twitch and couch. In-garden low hedges give a garden "edge" definition.

GRASSES

COUCH
KIKUYU
TWITCH
EXCLUDED

HIGH OUTER
HEDGE

COPROSMA
BANNAGRASS
WOODWORM
PAMPAS

WEED BARRIER

COMFREY
LEMONGRASS
CANNA LILY
GERANIUM
(Depending on local conditions + rampancy).

INNER GARDEN
LOW-HEDGES

ROSEMARY
TREE ONION
SUNROOT
(JERUSALEM ARTICHOKE

98

occasionally, forms a barrier between Zones I and II. Its berries are prized by chickens, and the leaves are an excellent source of potash, so it can be productively used in both Zones as a forage crop and as a rough mulch for garden plants. Canna lilies (*Canna edulis*) planted with lemongrass (*Cymbopogon citratus*) and comfrey (*Symphytum officinale*) form an impenetrable barrier for kikuyu grass in subtropical areas. Other successful barrier plants are wormwood, and autumn olive. In-garden hedges are smaller, often made up of rosemary and other perennial herbs and shrubs. There are excellent barrier hedge plants for every climate and condition.

In very windy areas such as on sea coasts, you can establish garden barriers immediately with a set of 3-5 tyres stacked in an arc against the wind (**Figure 5.5**). First newspaper and mulch the base of the tyre against weeds, then fill with earth, compost, scraps, hay, etc. and plant species able to withstand wind. The tyre arc not only blocks strong winds, but acts as a heat bank, protecting against frost and evening out temperature variations.

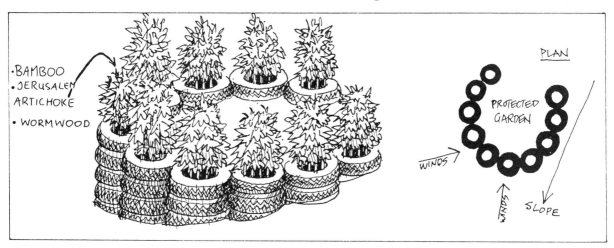

FIGURE 5.5 Hedge/windbreak using old tyres for a protected garden in very windy areas, e.g. sea coasts.

FIGURE 5.6 Vine arbour over garden path as a pergola.

99

■ VINE AND TRELLIS CROP

Using trellis to support both annual and perennial plants is the single most important space-saving device for both urban and rural gardens.

Trellis is placed against walls, fences, the carport, shed, shadehouse, verandah and patio; or can be specially-built as free-standing arbours (**Figure 5.6**), or even established over canals to provide shade for fish in hot climates. Trellis has a multitude of uses, including:

• Permanent barrier hedges around the garden (perennials such as passionfruit, hops, and vanilla).

• Deciduous shading of the house against summer sun (grape, wisteria).

• Permanent shading next to west walls (ivy, rambling roses).

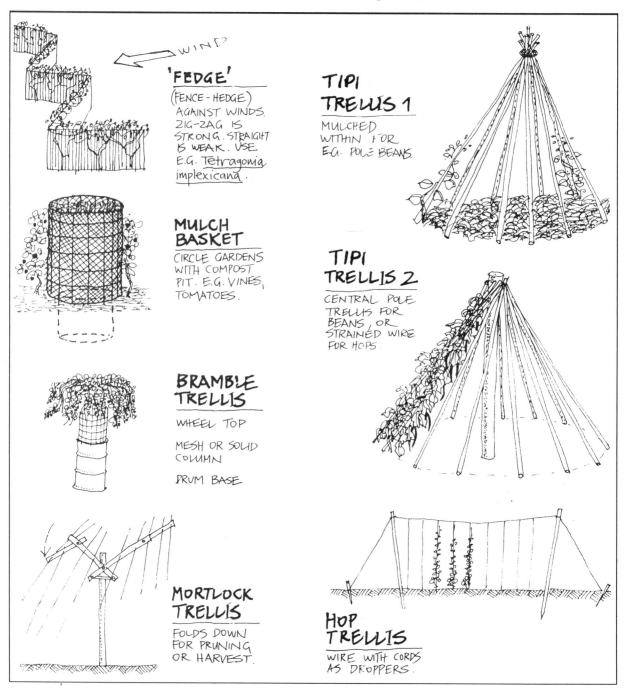

WIND

'FEDGE'
(FENCE - HEDGE)
AGAINST WINDS.
ZIG-ZAG IS
STRONG. STRAIGHT
IS WEAK. USE
E.G. Tetragonia
implexicana.

MULCH BASKET
CIRCLE GARDENS
WITH COMPOST
PIT. E.G. VINES,
TOMATOES.

BRAMBLE TRELLIS
WHEEL TOP
MESH OR SOLID COLUMN
DRUM BASE

MORTLOCK TRELLIS
FOLDS DOWN
FOR PRUNING
OR HARVEST.

TIPI TRELLIS 1
MULCHED
WITHIN FOR
E.G. POLE BEANS.

TIPI TRELLIS 2
CENTRAL POLE
TRELLIS FOR
BEANS, OR
STRAINED WIRE
FOR HOPS.

HOP TRELLIS
WIRE WITH CORDS
AS DROPPERS.

FIGURE 5.7 Trellis systems in the field or garden greatly enlarge cropping space.

• Summer playhouses (bean tipis) and living areas.

Figure 5.7 shows some trellis systems.

Sturdy trellis should be provided for all the vine plants, and care must be taken not to let rampant vines get out of control, especially in tropical and subtropical regions. Edible perennial vine plants include kiwifruit, passionfruit, grape, and hops. There are many other perennial, useful scramblers (flowers, leafy vegetation) providing shade and mulch material.

Annual vines include the cucumber, melon, and squash families, as well as the climbing legumes (beans, peas). Tomatoes (especially the cherry types) need to be treated as a vine, and can be staked or twined around mesh and string. In-garden trellis is provided for the smaller climbers, while the melon and squash vines are trained on outer fences, up arbours, or up onto the roof in urban areas. Be sure to provide a trellis structure consistent with the plant's climbing mechanism. **Figure 5.8** shows the different types of trellis supports used for different vine twining systems. Vines should be planted at frequent intervals for vertical growth.

■ GARDEN POND

A small garden pond used to grow water lilies or water chestnut is a haven for insect-eating frogs. Although such ponds are commercially-available from garden supply shops, they can be made from old tubs, plastic, or any non-leaky material.

Tyre Pond: An old truck or tractor tyre (not steel-belted!) is easily turned into a pond by cutting off one edge with a sharp knife. Dig a two-foot hole in the ground, wide enough to accommodate the width of the tyre, and tapering down (**Figure 5.9**). Line the hole with thick plastic, set the tyre on top of the plastic, and shovel dirt into the hole. Stones are set around the tyre to cover it, and a small, perennial flower such as alyssum is planted for decoration. Plant waterlily bulbs or water chestnuts into the earth at the bottom of the pond.

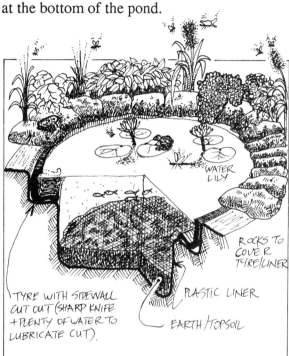

FIGURE 5.9 Garden tyre pond with water lilies, surrounding plants, frogs, insects, and fish.

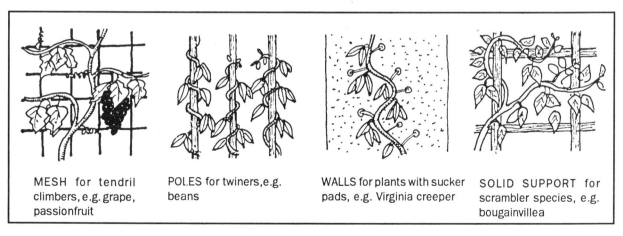

MESH for tendril climbers, e.g. grape, passionfruit

POLES for twiners, e.g. beans

WALLS for plants with sucker pads, e.g. Virginia creeper

SOLID SUPPORT for scrambler species, e.g. bougainvillea

FIGURE 5.8 TRELLIS SUPPORTS FOR DIFFERENT TYPES OF VINES.

101

■ SEEDLING BEDS AND NURSERY

Seedling beds should be close to hand in the garden, with easy path access. Earth from the seedling beds is always being taken away as vegetables are planted out and must be replaced from time to time. Or, raise seedlings in pots or trays containing a seed-raising mixture for easier handling from greenhouse to coldframe and out into the garden in appropriate weather conditions.

The nursery, an important feature in any initial permaculture, is placed where it will get plenty of water and attention. A greenhouse and shadehouse may be necessary in large-scale operations, but usually cold frames and a shade-cloth structure is all that is needed. Depending upon the scale of operations, the nursery is situated in Zone I or in Zone II, with consideration given to vehicle access (for nursery materials and perhaps sales), water, aspect, windbreaks, loading area, and so on.

Figure 5.10 shows an idealized Zone I layout for the temperate garden.

■ KEEPING ANNUALS PERENNIAL

In mild temperate climates, several techniques have been developed by gardeners to keep annuals in the garden "turning over". If a few leeks are left to run to seed, then dug up, many small bulbils can be found around the base of the stems. These are planted out in the same way as onion sets. Mature leeks cut at ground level (with the root left in the ground) will sprout again for another, smaller, harvest.

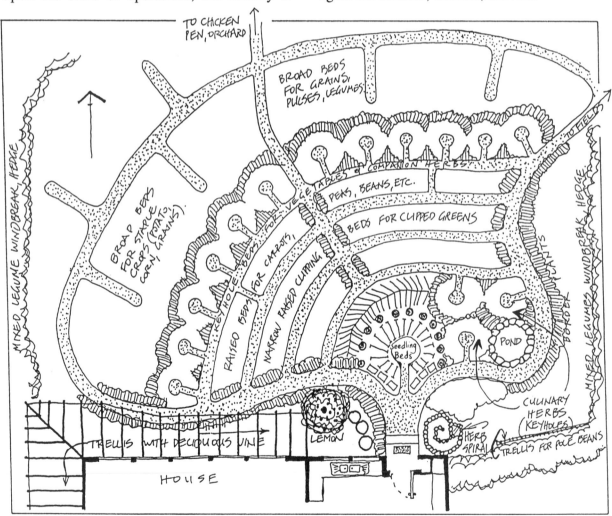

FIGURE 5.10 An idealised kitchen garden layout for temperate areas providing full nutrition, appropriate house climate, low maintenance compost area (near lemon tree), trellis crop, and herb spiral.

In the onion/leek group of plants, many are in any case perennial. Near the kitchen door we can plant two varieties of European chives (coarse- or fine-leaved), Asiatic garlic chives, and shallots of various types. Further away, as a border, set out potato onions (which give about 6-10 onions for every one planted), Welsh onions, evergreen bunching onions, the top bulbils of tree onions, and plant the cloves of garlic in the strawberry patch in autumn, or any space left in raised beds. Garlic bulbs, if allowed to multiply for two years, give a constant crop.

If the large pods at the base of broad bean plants are left on the ground to dry and are straw-mulched in late summer, they will resprout in autumn; or the crop may be pruned back hard after harvest and will sprout again. Seed potatoes left under mulch sprout in spring, and lettuce left to go to seed will scatter seedlings around their base for replanting. Parsley and many flat-seeded species re-seed freely in mulch, and their seedlings can be set out to grow. In fact, a small proportion (about 4-6%) of all crops sown can be let run to seed or ripen for scattering under mulch, rather than buying annual seed crop.

Various fruits and vegetables (tomatoes, pumpkin, melon), placed whole under mulch at harvest, ferment and rot, throwing up seedlings for new plantings. Carrot tops kept in a dark or cool place will sprout again, and can be set out to grow in soft soil (**Figure 5.11a**). Cabbages are cut low, and the stalk split crosswise with a knife. Smaller cabbage heads sprout, which are harvested in their turn, or divided up and replanted (**Figure 5.11b**).

In warm climates the axil shoots of tomatoes and related species can be pinched out and reset as small plants all summer (**Figure 5.11c**), the last lot potted and brought in to fruit over winter. Capsicum and chilies treated in this way may be winter pruned and then set outside in spring.

All these methods minimise resowing or making seed beds, and keep the garden turning over constantly.

THE INSTANT GARDEN

Sheet mulching for gardens is a technique which has been described by many people, with as many variations. It is my favourite technique as it gets you going immediately, without the back-breaking work of digging the soil for beds. You can start on almost any type of soil, except for those leached-out, rock-hard soils looking and feeling very much like concrete. With these, you build boxes *up off the ground* and cart in earth and compost materials to fill them.

Sheet mulching suppresses all weeds: ivy, onion and spear twitch, kikuyu and buffalo grass, docks, dandelions, oxalis, onion weed

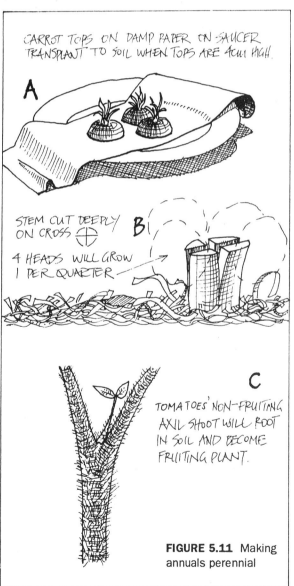

CARROT TOPS ON DAMP PAPER ON SAUCER. TRANSPLANT TO SOIL WHEN TOPS ARE 4cm HIGH.

A

STEM CUT DEEPLY ON CROSS ⊕ B
4 HEADS WILL GROW 1 PER QUARTER

C
TOMATOES' NON-FRUITING AXIL SHOOT WILL ROOT IN SOIL AND BECOME FRUITING PLANT.

FIGURE 5.11 Making annuals perennial

and even blackberries. The important thing is to fill up the area with plants, according to the prior planting plan you have worked out on paper, and to totally cover the area with mulch. For that reason, start with an area of about 4 square metres, and branch out as time and materials permit. Your first attempt should be very close to the house, preferably starting from a foundation or path which is itself weed-free. Thus, you are protected from an invasion of weeds from the rear. **Figure 5.12** shows the sequence for sheet mulching.

First, plant any large trees or shrubs. It is easier to plant these now than to dig through the mulch layer at a later date. Next, sprinkle the area with a bucket of dolomite (and gypsum, if the ground is particularly clayey), and chicken manure or blood and bone (to add nitrogen to start the process of reducing the carbon in the following layers). A bucket or two of compost scraps can also be scattered, for the worms. If you have a source of weed-seedy hay or like material, place this also over the area.

Don't bother to dig, level, or weed. Now, proceed to tile and overlap the area with sheet mulch material. This can be cardboard, wallboard, newspaper, old carpet (non-synthetic), underfelt and anything that will eventually break down and provide nutrients for plants. Cover the area completely, leaving no holes for weeds to poke through. If you have a valuable tree or shrub in the way, tear paper halfway across and pull it around the stem. Serve another, at right-angles to the first. Go on, leaving only valuable plants with their stems and leaves poking out.

Water this layer well; it will start the processes going. Then apply a 7.5cm layer of either (or mixed) horse-stable straw; poultry manure in sawdust; leaf mould or raked leaves; seagrass or seaweed.

All of these contain essential elements, and hold water well. Follow these with dry, weed-seed-free material on top, of at least 15cm of pine or casuarina needles; rice husks; nut shells; cocoa bean husks; leaf mould or raked leaves; seagrass; dry straw (*not* hay); bark, chips, or sawdust or any of these mixed.

Water until fairly well soaked. Now, take *large* seeds (beans, peas), tubers (potato, sunroot), small plants (herbs, tomato, celery, lettuce, cabbage) and small potted plants. Set them out as follows:

With your hand, burrow down a small hole to the base of the loose top mulch. Punch or slit a hole in the paper, carpet, etc. with an old axe or knife. Place a double handful of earth in this hole, and push in the seed or tuber, or plant the small seedling in it. For seeds and tubers, pull the mulch back over. For seedlings, hold the leaves softly in one hand, and bring the mulch up to the *base* of the plant.

If you must use small seed, do it this way: Pull back the mulch in a row; lay down a line of sand, or fine soil, and sow small seeds of radish, carrot, etc. Water, and cover with a narrow board for a few days, or until the seeds have sprouted (or sprout them first on damp paper). Then remove the board and draw mulch up as the tops grow.

Root crops do not do well in the first year, as the soil below is still compacted and there may be too much manure. Plant daikon radish, whose 30-60 cm root will begin to break up the compacted ground. Plant most root crops in the second year (or dig a separate bed for them), when it is only necessary to pull back the loose top mulch to reveal a layer of fine dark soil.

By the end of the first summer, the soil is revolutionised, and will contain hundreds of worms and soil bacteria. Just add a little top mulch to keep levels up, usually a mix of chips, bark, pine needles, and hay. Scatter some lime or blood and bone. Annual plants need occasional fresh mulch after harvest; their outer leaves are "tucked under" the mulch layer, as are all your food wastes from the kitchen. Worms are so active that the leaves and peelings disappear overnight. Leather boots take a little longer, old jeans a week or so, and dead ducks a few days.

In the first year, you need to water fairly frequently, as the layer of fungal hyphae and plants at the base of the mulch are slow to develop. As in normal gardening, all newly-

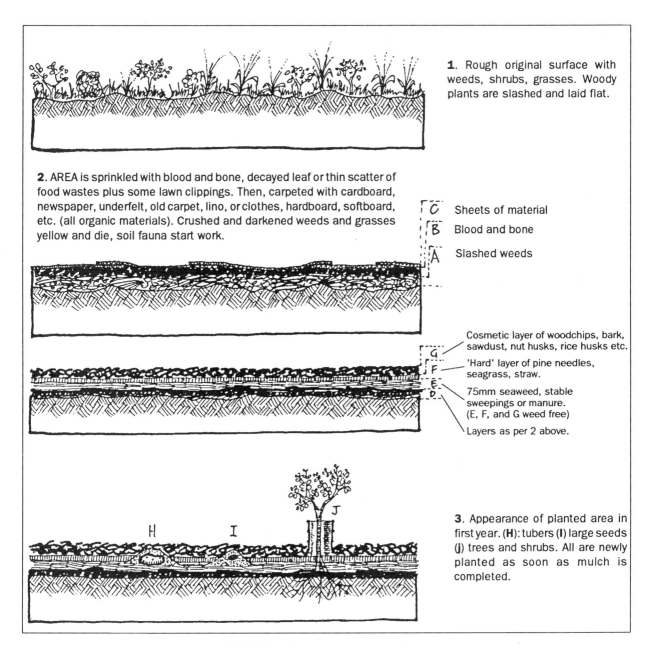

1. Rough original surface with weeds, shrubs, grasses. Woody plants are slashed and laid flat.

2. AREA is sprinkled with blood and bone, decayed leaf or thin scatter of food wastes plus some lawn clippings. Then, carpeted with cardboard, newspaper, underfelt, old carpet, lino, or clothes, hardboard, softboard, etc. (all organic materials). Crushed and darkened weeds and grasses yellow and die, soil fauna start work.

C — Sheets of material
B — Blood and bone
A — Slashed weeds

Cosmetic layer of woodchips, bark, sawdust, nut husks, rice husks etc.
'Hard' layer of pine needles, seagrass, straw.
75mm seaweed, stable sweepings or manure. (E, F, and G weed free)
Layers as per 2 above.

3. Appearance of planted area in first year. (**H**): tubers (**I**) large seeds (**J**) trees and shrubs. All are newly planted as soon as mulch is completed.

FIGURE 5.12 Steps in sheet mulching.

planted seedlings need water initially.

There is no need to rotate plants in this system, or to rest the ground. Potatoes are simply placed on top of the old mulch, and re-mulched. There is no need to leave room to hoe or dig either, so plants may be stacked much more closely, and preferably in mixed beds rather than in strict rows. By frequent and random replanting, the garden will start to assume the healthy appearance of a mixed herbal pasture. This diversity of plants act as hosts for a range of insects, frogs, and birds and is a major factor in successful pest control.

Some strong weeds may force through. Push the weed down in the mulch, put damp paper on its head, cover with sawdust. If 10% of the kikuyu or twitch comes up, sheet with paper and cover with mulch. All eventually die out under this treatment, leaving the area clear of weeds; only your plants have their heads in the air. Another ploy is to dig up dock roots, bury kitchen scraps there, and re-mulch.

Never bury sawdust or woodchips; just put them on top where atmospheric nitrogen breaks down the wood. Worms add sufficient manure to supply the base manure. Keep the mulch loose, don't let it mat, and thus mix lawn clippings or sawdust with stiff dry material like chips or pine needles, bark, etc.

THE URBAN AND SUBURBAN PERMACULTURE GARDEN

Urban/suburban design takes the same principles of permaculture and applies it to a smaller scale. Usually, there is space for only Zone I and some Zone II plants, animals, and structures. The important thing to remember is that the smaller the available space, the greater care that must be taken to both intensify food production, and to minimise waste space by using spiral, keyhole, trellis, least-path systems, and stacked or clumped plantings.

■ SMALL URBAN SPACE

This situation requires most thought, but it is surprising how much food can be grown on window-sills, roofs, verandahs, narrow walkways, and patios. Plants can even be grown indoors in pots as long as they are wheeled out to a sunny location; most plants need at least 6 hours of sunlight a day during the growing season.

Containers can be of almost anything: plastic garden pots, wastepaper bins, old baskets, half-filled sacks, toy boxes. Poke holes in them so that water can escape, and be sure their combined weight does not bring the balcony crashing down onto the people beneath. A light soil mixture is made up specially for container planting on balconies and roofs; it may need more frequent watering.

Deeper containers are needed for root vegetables. Potatoes are grown in a small area through the use of a potato box, which is made from a 44 gallon drum, a wooden box, or (outdoors) old railway sleepers or car tyres. Potatoes are placed on a bed of mulch inside the box, with mulch put over them. As the potatoes sprout and grow, more mulch is piled in, until the green leafy tops are sticking up out of the box. In this way, potatoes are formed from the covered stem and are more easily picked than if grown in hard ground (**Figure 5.13**).

Choose plants you are certain to eat, which are particularly nutritious, and which can be picked at least twice a week, such as capsicums (bell peppers), tomatoes, parsley, chives, silver beet (Swiss chard), and lettuce. If space is limited, stick to herbs that are frequently used (thyme, marjoram, basil).

Window-sill space is better used if hanging baskets or 2-3 shelves are added (**Figure 5.14**). Better still is a window-box greenhouse

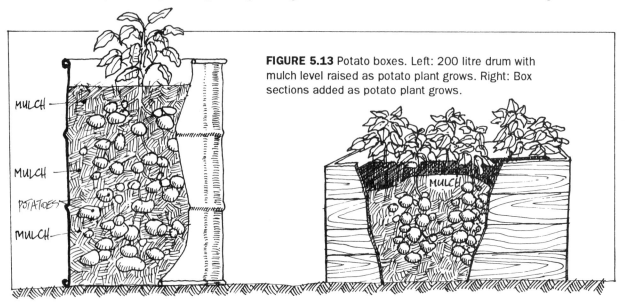

FIGURE 5.13 Potato boxes. Left: 200 litre drum with mulch level raised as potato plant grows. Right: Box sections added as potato plant grows.

set out from the wall, facing the sun, as illustrated in Chapter 4 (**Figure 4.10**).

On verandahs and small patios, plants should be in tiered plantings with taller plants at the back so as not to shade smaller species. Two or three shelves of pots or long planter boxes can be stacked vertically (**Figure 5.15a**).

Other well-known ways to grow food in a small space include sprouting alfalfa, sunflower, and mung seeds, and growing a sack or two of mushrooms in a cool dark place.

Kitchen scraps are composted in a two-bucket system under the sink, adding garden trimmings to food. Some scraps, such as orange peels and uncrushed eggshells, take a long time to break down, but this is easily achieved if you take the time to cut and crush them.

For apartment dwellers, trellis is best trained around the verandah/balcony or set up against walls outside the window (**Figure 5.15a and c**).

■ SUBURBAN BLOCKS

Most people in Australia own or rent a house with a small to medium-sized front and back

FIGURE 5.14 Salad greens in hanging baskets and on windowsills for apartment dwellers.

yard. Many of these houses could accommodate a small greenhouse or shadehouse, trellis systems, fruit trees, a polyculture of annual and perennial plants, and some small, quiet livestock such as duck, quail, bees, and bantam chickens. See **Figure 5.16** for an idealised "before" and "after" view of a typical suburban block.

Trellis takes the place of shade trees, many of which are too large for urban blocks. Always be careful to design the trellis systems so that they do not shade out ground beds of smaller plants, unless those plants benefit under shade.

Fruit Trees: Miniature fruit trees, which are grown either in the ground or in large pots, are compact (usually only 2 metres tall at maturity) and bear normal-sized fruits within a few years. Their disadvantages are initial cost, more care, and a shorter lifespan.

Grafted trees are also very valuable in a small garden. Branches of one variety of apple, for example, can be grafted onto another variety to ensure cross-pollination or fruits that ripen at different times. Better still, it is possible to graft three or more types of fruit onto one tree. A peach tree, for example, can bear almonds, nectarines, apricots, and Japanese and European plums. Apples, cherries and pears will not grow on peach but any one of these can be grafted to support different varieties of that particular species.

Always consider the height and spread of trees, as they may eventually shade out the garden. Almost all fruit trees can be pruned and trained against a wall or fence (espalier). Although it requires careful pruning and tying, advantages are easier picking, netting against birds, and saving space.

Garden Beds: Any sort of garden bed can be used, from mounded, sunken, keyhole, circle, to earth- and compost-filled boxes. One technique on hard ground or rubble is to build compost-filled circle beds. The main advantages of these round beds are:

• Water savings: A circle is watered by one sprinkler more efficiently than a long row of vegetables.

• Nutrient concentration: the circle is a "dumping ground" for all kitchen scraps, vegetable trimmings, manures, and other added organics, forming a rich area of compost and humus.

• Circle gardens can be constructed in difficult climates (particularly arid regions) and in places where the ground is unsuitable for growing, e.g. rubble, hardpan, sand, and clay, as they grow entirely in soil that has been collected from the area or composted on site.

To build an above-ground circle bed (**Figure 5.17a**), proceed as follows:

1. If possible, dig a circular hole in the ground a little bigger than the circumference of the circle. The diameter should be as far as you can reach to the centre from any point, say 1.2

FIGURE 5.15 (A) Cut-away view of a patio set up for herb, vegetable, and small fruit in beds and pots **(B)** Outside window box and trellis **(C)** Verandah trellis for shade and fruit.

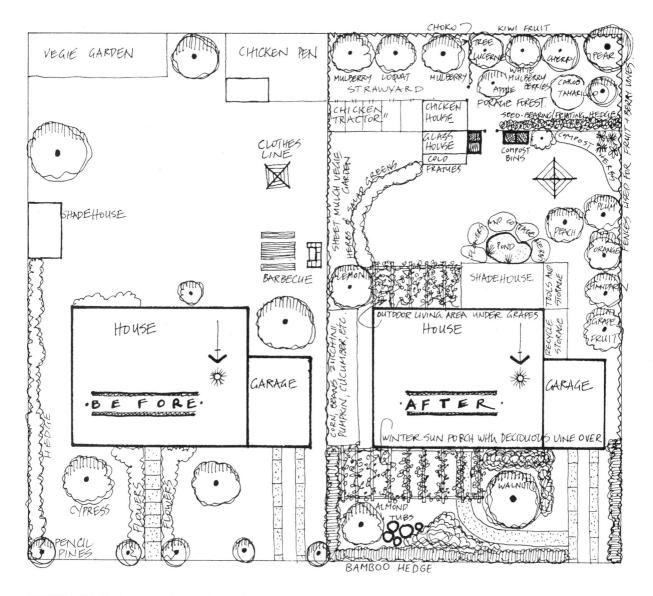

FIGURE 5.16 Before and after versions of a suburban block. BEFORE: High maintenance, low yield. AFTER: Low maintenance, high yield.(Adapted from a drawing by Robyn Francis: *Chickens in a Permaculture Garden.*

metres across in total. The depth is one shovel deep, with the earth put aside (on a canvas or plastic sheet). The bottom of the hole is turned over or loosened.

2. Place a 60cm high circle of chicken wire around the hole. Throw earth around the edge of the wire to secure it in place. To keep earth and other fine material from oozing out of the wire, use straw as a barrier next to the wire. As material is put into the circle, it will bulge but remain taut.

3. Start filling the hole with food scraps, compost, leaves, twigs, etc. in layers with the earth you had taken out before. From time to time, sprinkle in nutrients: cow manure, aged chicken manure, some form of phosphate, a sprinkling of ashes, lime, blood and bone, seaweed, etc.

4. Build it to the top of the chicken wire, and sprinkle with a layer of fine earth.

The actual growing is done in this small space, but the plants use up a bigger area because they can spread out from the circle. Cucumbers and zucchini drape over the bed and trail off onto the ground, while tomatoes are staked up outside the circle.

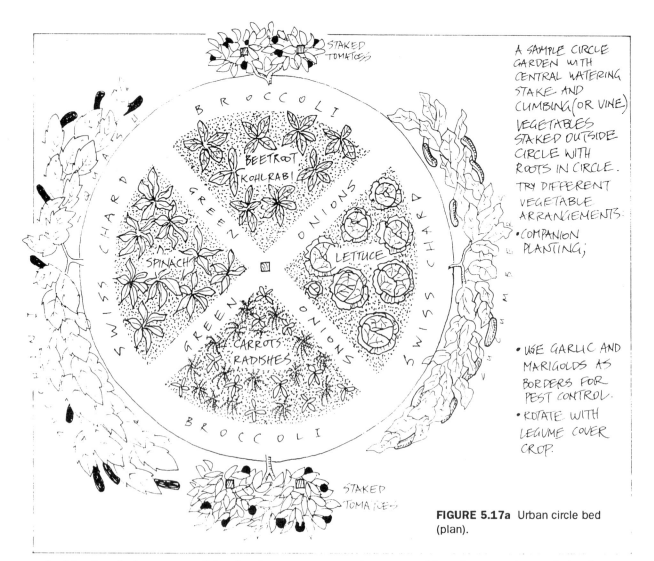

Within the figure:

STAKED TOMATOES

A SAMPLE CIRCLE GARDEN WITH CENTRAL WATERING STAKE AND CLIMBING (OR VINE) VEGETABLES STAKED OUTSIDE CIRCLE WITH ROOTS IN CIRCLE. TRY DIFFERENT VEGETABLE ARRANGEMENTS:

• COMPANION PLANTING;

• USE GARLIC AND MARIGOLDS AS BORDERS FOR PEST CONTROL.

• ROTATE WITH LEGUME COVER CROP.

BROCCOLI

SWISS CHARD

BEETROOT KOHLRABI

GREEN ONIONS

SPINACH

GREEN ONIONS

LETTUCE

SWISS CHARD

CARROTS RADISHES

BROCCOLI

STAKED TOMATOES

FIGURE 5.17a Urban circle bed (plan).

Inside the circle any sensible planting combination can be followed, particularly of planting a fast-growing crop with a slow-growing one (carrots, shallots and radishes; broccoli and lettuce) as one is removed while the other is still growing. Care must be taken in the winter garden not to shade out small plants with taller-growing species; this is not such a problem in the summer garden when the sun is overhead.

As plants are harvested, others are put in their place if there is enough light. (With enough water and nutrients, the only limitation is light.) Three beds will keep three people in salad and other vegetables all year, and once set up need little attention.

Watering is easy, as a sprinkler is placed at the top of a stake in the middle of a circle, or a tube of spray emitters from a drip irrigation system is tied to poles. For early raising of spring vegetables, drape a sheet of plastic over the stake and around the circle, leaving a small opening around the base for air circulation (**Figure 5.17b**).

Added to the circle gardens and trellis systems, a flattish roof can be used to grow pumpkins and watermelons. If you have a wooden fence next to the house, construct a column of black plastic (*not* clear plastic—the roots will burn up) and chicken wire in the corner (**Figure 5.17c**), nailing the chicken wire into the fence. Fill the column with nutrient-rich earth and plant seeds. As the seedlings grow, clip off all but two strong stalks for each plant and lead them to the roof, where they can spread freely. The important thing to remember is to water frequently as the column dries out fairly quickly; it is best to have a drip system operating on automatic, if possible.

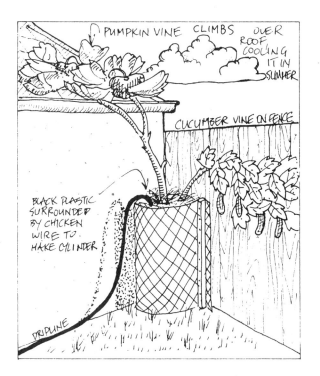

FIGURE 5.17b Circle garden with plastic tent.

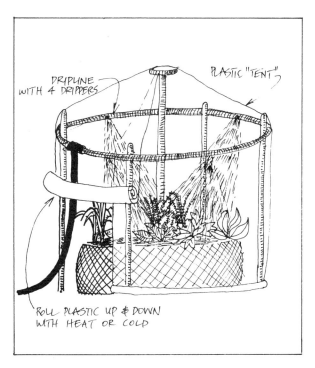

FIGURE 5.17c Chicken-wire column for rooftop vegetables.

The Suburban Lawn

The American lawn uses more resources than any other agricultural industry in the world. It uses more phosphates than India, and puts on more poisons than any other form of agriculture. The American lawn could feed continents if people had more social responsibility. If we put the same amount of manpower, fuel and energy into reforestation we could reforest the entire continent. A house with two cars, a dog, and a lawn uses more resources and energy than a village of 2000 Africans.

Often you will see a little house on a residential block, surrounded by flowers and lawn and perhaps a bit of shrubbery. Behind the house, way in the back and maybe hidden by a discreet trellis, will be a small vegetable garden. You recognise the pattern. It is so universal that to move a cabbage up on to this lawn is a cause for total neighbourhood consternation. My favourite story is that of a man in Tasmania who dared plant cabbages on his "nature strip"—that sacred and formal grassy area between sidewalk and street. Having thus demonstrated his total lack of the sense of fitness of things, he was sharply reminded of his error when the local council sent trunks and men to uproot the vegetables (which were merely useful, and therefore of no aesthetic value). I must, in all fairness, say this occurred in 1977 and by 1979 the council had tentatively begun to plant fruit and nut trees in their public parks.

Yet why should it be indecent to have anything useful in the front half of your property or around the house where people can see it? Why is it low-status to make that area productive? The condition is peculiar to the British landscaping ethic; what we are really looking at here is a miniature British country estate, designed for people who had servants. The tradition has moved right into the cities, and right down to quarter acre patches. It has become a cultural status symbol to present a non-productive facade. The lawn and its shrubbery is a forcing of nature and landscape into a salute to wealth and power, and has no other purpose or function.

The only thing that such designs demonstrate is that power can force men and women to waste their energies in controlled, menial and meaningless toil. The lawn gardener is a schizoid serf as well as the feudal lord, following his lawnmower and wielding his hedge clippers, and contorting roses and privet into fanciful and meaningless topiary.

If you've inherited a large lawn, never fear: help is at hand! It is easily turned into productive space in a few hours by sheet mulching with newspaper and straw (depending on family needs, a small space might be saved as a children's play area), and can be designed to be both aesthetically-pleasing *and* productive by planting:

• *Shrubs*: Gooseberries, blueberries, currants, rhubarb.

• *Flowers for salads*: borage, nasturtium, calendula, daylily (for a list of edible flowers, see Appendix B).

• *Herbs*: thyme, lavender, rosemary, oregano, marjoram.

• *Colourful vegetables*: variegated kale, chili peppers, capsicum (red, green, yellow), egg-plant (elongated, black, yellow), yard-long cucumbers, watermelons, squash on trellis, scarlet runner beans (beautiful flowers), cherry tomatoes, asparagus, pumpkin.

• *Carpeting plants*: chamomile, alpine strawberries.

• *Trees*: citrus, persimmon (orange fruits hang off leafless trees in autumn), almond and apricot (pink and white flowers in spring).

Thus, an energy-consuming, unproductive lawn is turned into a large food-producing area containing 100-200 plant species in less than six months. If all suburban lawns were so transformed, urban food needs could be cut by at least 20%.

5.5
COLD AREA GARDEN DESIGN

The main design considerations in cold areas are in extending the growing season through the use of plastic or glass; protecting plants from frost for as long as possible; using locally-adapted shrubs and trees for wind-break, mulch, and fodder; growing vegetable varieties which are specially developed for short-season growing; and storing fruits and vegetables in autumn for winter use.

Vegetable gardens should be sited close to the house for easy access and so that plants can be quickly covered on frosty nights. If the garden is on a sloping site, ensure that cold air is allowed to drain downhill and that there are no barriers such as a dense hedge or wall which can act to dam the cold air. Try to provide a break or pathway through this barrier to allow cold air to flow downslope. In mild frost areas, raised beds can save plants from ground frosts.

The single most important garden/house structure is a well-insulated greenhouse, where the floor inside the walls is insulated from the cold earth outside. Heat masses are a bank of 44 gallon drums filled with water or even large plastic tanks serving as fish ponds, a successful strategy used by the New Alchemy Institute in Massachusetts in their large bio-shelter. (**Figure 5.18**)

The areas below plant benches, if used to house rabbits, guinea pigs, poultry, or any small domestic animals at night, will provide considerable winter heat (see also Chapter 6 for a chicken-heated greenhouse design). Insulated boxes of actively "cooking" compost located inside or just outside the greenhouse will provide warmth, as will hot water pipes and storages filled from solar heat collectors. Even water pipes connected from the shower can be used under growing beds, after first going through a filter.

Another style of greenhouse, particularly suited for inner-city dwellers, was designed by Dr Sonja Wallman for intensive production of food in a densely built-up area of Berlin. An enthusiastic gardener, Dr Wallman has also developed similar greenhouses in cold climate areas of New Hampshire, in the USA.

This greenhouse differs from other models in that, even in the cold continental climate of Berlin, it needs no additional heating.

This is achieved by the following design principles:

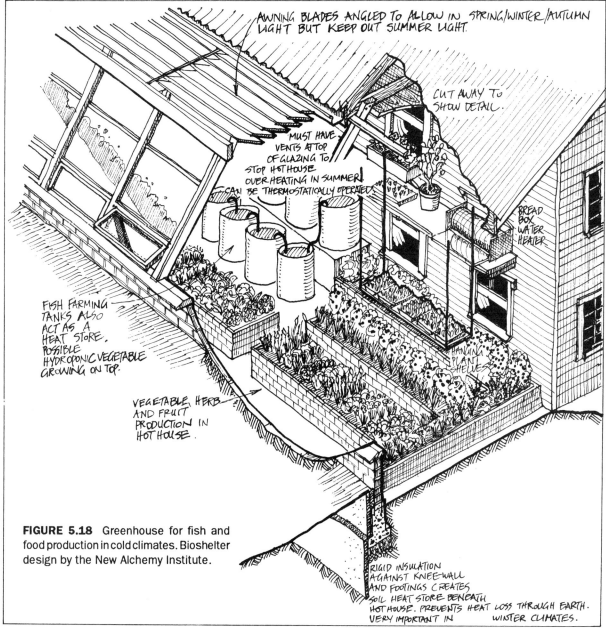

AWNING BLADES ANGLED TO ALLOW IN SPRING/WINTER/AUTUMN LIGHT BUT KEEP OUT SUMMER LIGHT.

CUT AWAY TO SHOW DETAIL.

MUST HAVE VENTS AT TOP OF GLAZING TO STOP HOT HOUSE OVER HEATING IN SUMMER! CAN BE THERMOSTATICALLY OPERATED

BREAD BOX WATER HEATER

FISH FARMING TANKS ALSO ACT AS A HEAT STORE. POSSIBLE HYDROPONIC VEGETABLE GROWING ON TOP.

HANGING PLANT SHELVES

VEGETABLE, HERB AND FRUIT PRODUCTION IN HOT HOUSE.

RIGID INSULATION AGAINST KNEE WALL AND FOOTINGS CREATES SOIL HEAT STORE BENEATH HOT HOUSE. PREVENTS HEAT LOSS THROUGH EARTH. VERY IMPORTANT IN WINTER CLIMATES.

FIGURE 5.18 Greenhouse for fish and food production in cold climates. Bioshelter design by the New Alchemy Institute.

• The greenhouse is not free-standing but attached to an existing house.

• Orientation to the sun-sector (south-east to south-west) following the exact sun-angles of summer and winter.

• The wall of the house and the use of double-sheeted glass provide extensive heat insulation. Thus the greenhouse is even able to conserve energy as it acts as a suntrap and buffer-zone.

• It also acts as an airfilter, improving air quality throughout the house, an important consideration in areas of heavy traffic pollution.

In winter, the sun heats the rear wall of the house which serves as a heat storage. The heat thus collected during the day radiates into the house during the evening and thus helps to save on domestic energy costs, during the average 250 days per year when temperate houses require heating. In summer, the insulated solid part of the sloping roof protects the rear wall from direct sun rays. Ventilation flaps, arranged in the greenhouse and rear wall, direct air flow.

By overlapping plant species and size, as well as harvesting methods, (ie plucking outer leaves of lettuce rather than the whole plant), it

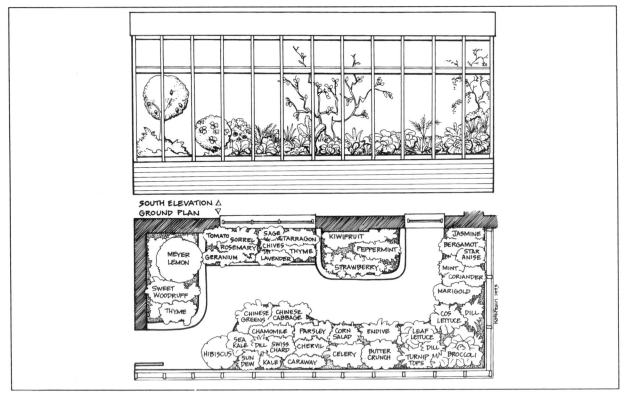

FIGURE 5.18a Greenhouse plan producing 70% of the salad needs of typical family.

is possible to produce within the 20m² greenhouse **(Figure 5.18a)**, approximately 70% of the needs of fruit and salads for a family of three to four persons.

Organic waste from the kitchen, household and greenhouse is turned into high-value compost by worms (in a wormbox). Together with mulch, the soil is regenerated continuously. Along the wall of the house are planted herbs, while along the outer glass wall are different cabbages and lettuces. When the soil temperature reaches 23° C, the summer planting with annuals can follow. The winter plants will then be replaced one by one by tomatoes (on trellis), cucumbers, pole beans, nasturtiums, basil, etc. The hardy or perennial plants remain in their place.

Raised insulated beds are constructed of brick. The height of 80 cm allows working and harvesting without backaches.

Work input for the greenhouse is estimated to be one weekend each for the summer and the winter planting respectively. The need for upkeep and watering is about 15-20 minutes daily. However, this is more than compensated for by the savings of money and time in not having to shop for fruit and vegetables every day.

See Box opposite for a list of suitable plants for the inner-city greenhouse.

Other "mini-greenhouse" devices that have been used by cold area gardeners are cloches, inverted glass jugs, and moveable plastic frames in various shapes. **(Figure 5.19)**.

Rock walls backed by such reflective trees as birch give an early warm site to plant out vegetables. Stone walls in a gentle arc form warm early growing sites, as do semicircles of tyres facing into the low sun. Such embayments can be plastic or glass-covered to assist heat retention, or piles of tyres can be topped with glass as miniature grow-holes, especially if the tyres are earth-filled to retain daytime heat. The Chinese use slanted bamboo and straw lean-tos to achieve this early growth of vegetables and to extend their growing season. The shade side of such shelters accumulate snow for insulation.

Vegetables that withstand most frosts are some root crops (carrot, leeks, turnips); these must be covered with bales of hay to keep the ground from freezing. It is best to group such plants together, although the entire garden will benefit from a thick layer of hay in winter. Kale also withstands winter frosts. Many vegetables

Plants for a Solar Heated Greenhouse
Dr Sonja Wallman

Annual Vegetables, Herbs & Edible Flowers

Oriental Cabbage Species
(hardy, fast growing cabbage and greens)

Hon Tsai Tai (red-purple flowering stem with dark green leaves)
Delicate Green (dark green leaves, tasting mild, similar to spinach)
Chinese Kale (edible green leaves, flowers smell of roses)
Chinese Broccoli (like broccoli, flowers stems and leaves are edible)
Kyona Mijuna (biennial Japanese mustard, tasting mild, yielding well under extreme temperature variations)

Turnip Tops	Broccoli	Okra
Butte Cabbage	Lamb's Lettuce	Sorrel
Winter Endive	Leaf Lettuce	Corn Salad
Swiss Chard	Japanese Greens	Cos

Annual Herbs
Coriander (edible herb which is important for the lifecycle of ladybirds)
Dark Opal Basil (a kind of sweet basil with dark purple leaves which is more hardy than the better known basil)
Cucumber (female cucumber without pips which is trained up trellises in green houses: an example of this kind - 'Sandra')

Aniseed	Camomile	Chervil
DellSoup	Celery	Caraway
Italian Parsley		

Edible Flowers
Marigold	Nasturtium

Hardy Edible Fruit, Herbs & Flowers

Fruit
Alpine Strawberry
Chinese Gooseberry (kiwi - a vinelike, climbing cordon plant with sweet fruit: it is necessary to grow male and female plants to ensure fruiting)
Citrus Fruit (Meyer's lemon, Persian lime & Calamondin orange are all kinds suitable for a greenhouse; these kinds produce sweet smelling flowers and edible fruit, throughout the year)
Tomatoes (Sweet 100 Cherry tomatoes grow and have fruit for several years in a greenhouse, if the temperature doesn't sink below zero)

Herbs
Spearmint	Peppermint	Bergamot
Sage	Sorrel	Mugwort
Chives	Woodruff	Thyme
Tarragon	Rosemary	

Flowers
Jasmin (tea)

Special Plants for Insects
Pelargonium (host plant of the wasp *Encarzia formosa* - parasite of whitefly)
Sun-Dew (*Drosera* - Insect eating plant which catches small flies; attractive plant; grows well)

Aromatic Plants of Non-Edible Flowers
Lavender (infusion for nervous ailments as bath additive, externally for rheumatism, bruises, sciatica, neuralgic pains)
Jasmine   Hibiscus

---

can be harvested in autumn and kept clean and dry in cellars; these are often layered in sand (carrots) or wrapped individually in newspaper (tomatoes). Tomato plants may also be pulled whole from the ground and hung upside down in the cellar; tomatoes will then slowly ripen.

A look around the district will reveal useful hedge, windbreak, mulch, and animal forage species suited to the climate. There are many cold area fruit varieties of apple, quince, blueberry, rosehip, grape, persimmon, and even a hardy kiwifruit (Actinidia arguta). Nuts include walnut and American chestnut. Animal forages are honey locust, oak (acorns), and autumn olive. The Seed Savers Exchanges, both in the USA and Australia, carry a fascinating range of open-pollinated and heritage seeds, including many specially adapted for cold climate conditions. Contact details are included in Appendix E: Permaculture Resources.

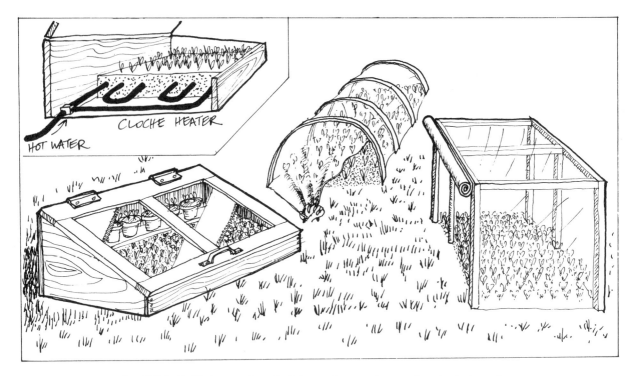

FIGURE 5.19 Various styles of mini-greenhouses for starting plants in spring.

5.6
TROPICAL GARDENS

Like temperate gardens, the tropical garden needs a variety of perennials, annuals, vine crop, and barrier hedges. In addition, it contains papaya (pawpaw) and nitrogen-fixing thin-foliage trees as a canopy above the garden to give shade from the sun.

Tropical soils are thin and leached out due to heavy rains, so it is essential to interplant leguminous green crops (both perennial and annual) within the garden as a cut-and-mulch system. Mulch can be cut all year from a variety of non-legume hedge and understorey. Such species as *Nicotiana*, wild ginger, lemongrass, bamboo (leaves), vetiver grass, and crop wastes from maize, sesbania, and soft ground legumes or comfrey provide constant mulch so that the coppicing of susceptible tree legumes is reduced. All garden wastes are returned to the beds, and beds are replanted as they are harvested. A top mulch of straw, bark, dry manure, or woodchips is added annually, or whenever needed.

■ GARDEN BEDS

Garden beds should be mounded to shed water, particularly in the wet season; otherwise they will become waterlogged and plants will rot. There are various possible bed shapes. (**Figure 5.20**), depending on climate. Briefly, earth *mounding* is best for humid tropics while sunken beds are best for dry tropics.

Ridges. Ridges of 0.5m x 1m increase yields in cassava, sweet potato, potato, and yam crop. Mulch and green crop can be grown between the ridges. Pineapple and ginger also prefer ridges in wet areas. Leucaena intercrop for mulch is on mounds, while maize and green mulch (beans) occupy hollows. Ridges permit deep mulching for low crop such as pineapple, with mulch being applied between the ridges.

Basins: Even shallow basins, aid dryland taro and banana, or patches of Chinese water chestnut. Soil is more easily saturated, and deep mulch keeps it from drying out.

Boxes made from palm trunks are ideal mulch-holders for yams, banana, vanilla orchids, vines generally, and borders of beds in home gardens.

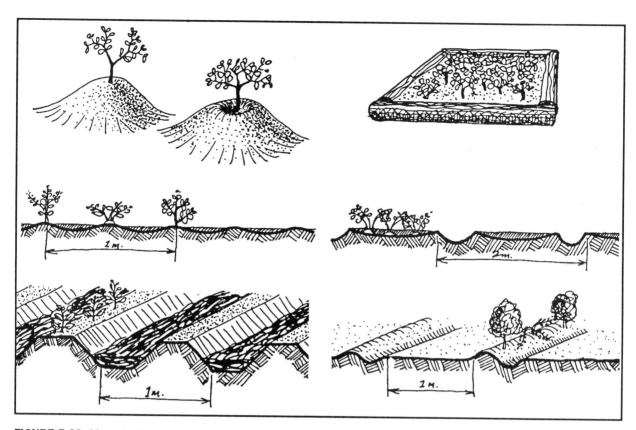

FIGURE 5.20 Mounds, boxes, ridges and basins are some appropriate garden bed shapes for tropical climates.

Cut palm trunks are also useful for holding earth for terrace beds across medium slopes.

■ BANANA/PAPAYA CIRCLE

A wet, mulched circle surrounded by bananas, papayas (pawpaws) and sweet potato is a useful area to compost scraps, to accommodate excess runoff, or to contain an outdoor shower (**Figure 5.21**).

Steps in the process are:

1. Describe a circle 2 metres across and dig the topsoil (or subsoil) to a dish shape, ridges on the outside, and about 0.6-1 metre deep from hollow to rim. A narrow inlet at ground level can be dug to accept rainwater runoff.

2. Cover the circle with wet paper or cardboard, banana leaves, or any mulch material such as coarse twigs, hay, rice husks, etc. Add manures, ash, lime, dolomite, or other fertilisers. Building these materials up in layers of 15-20cm, overfill the circle so that it bulges up at the top (it will soon sink down). If stones are

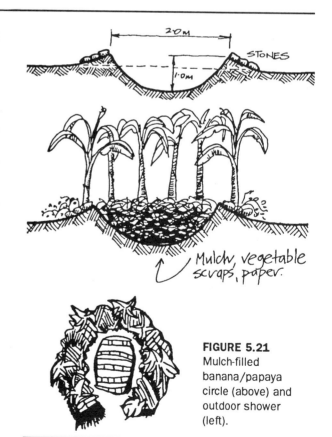

FIGURE 5.21
Mulch-filled banana/papaya circle (above) and outdoor shower (left).

117

available, bank them to the outside of the rim.

3. Plant the rim to 4-5 papaya (a tall variety), 4 bananas (dwarf types), and 8-10 sweet potatoes. Yams or taro can be planted inside the rim, *or* a wooden, slatted platform placed inside for an outside shower.

■ WEED BARRIERS AND MULCH SUPPLY

Because of the prolific growth in the tropics, weeds are often a problem. Around the annual, mulched gardens, a band of grass-barrier plants prevents weed re-invasion.

A combination of the following usually works:
• a deep-rooted broadleaf (comfrey);
• a clump grass which does not seed down or is not browsed (lemongrass, vetiver grass);
• a carpeting plant such as sweet potato; and
• a bulb such as *Canna edulis*.

Bordering a garden, woody legumes such as moringa (horse-radish or drumstick tree), sesbania, leucaena, calliandra, and sunn hemp (*Crotalaria*) provide mulch for the garden beds and fodder to domestic livestock. Behind that, a taller border of cassava, banana, papaya, pigeon pea, and leucaena forms a hedge or windbreak.

To discourage animals, thorny or inedible hedges are planted around the garden. Plants that make good live fences are: cassava, cactus, hibiscus, bamboo, and a double row of spiny pineapples.

■ TROPICAL POLYCULTURE

As usual, diversity of garden species works best.

The following are some common planting arrangements found in Southeast Asian home gardens (from *The UNICEF Home Gardens Handbook*, P. Somers):
• Multi-storey tree crops: top layer of coconut, with middle layer of jakfruit and avocado. Next layer of banana, papaya and coffee, under which are planted winged bean and other edible vines growing on the tree trunks. Lowest layer: pineapple and taro.
• Climbing legumes: yardlong beans, winged bean and lima beans planted to one leucaena

stake or untrimmed piece of bamboo.
• Circle plantings: banana growing in the middle surrounded by cassava and tomato; winged bean growing on the banana; sweet potato as a ground cover. Mushrooms growing inside the hill of bananas.
• Water canal from the kitchen/shower feeding banana, sugarcane, kang kong, and taro.
• Trellis over an irrigation canal: bittermelon, squash, climbing legumes.

When planting trees in the garden, or close to each other, it is important to know their characteristics, such as height of mature trees, fruiting habits (plant a tree that fruits on the outside branches next to one that fruits on the inside to minimise light competition), drought-resistance, and shape. Generally, small trees with open foliage are the best to plant near the annual garden, with trees gradually getting larger towards the edge and inside of Zone II.

While a complex polyculture of many hundreds of species delights both the naturalist and the householder, it becomes difficult to control an *extensive* rich polyculture and collect its products. Very complex polycultures work best at a small scale and with close attention from people.

■ TROPICAL GARDEN PROBLEMS

Problems are numerous in tropical gardens, especially insect and rodent pests, wild pig, snails, and sometimes monkeys and larger animals. Thus there is a need for spiny or woven fences of *Euphorbia*, yatay palm, bamboo.

By planting a mixed, multi-storey system, insect pest problems are minimised; frogs, spiders, small insectivorous birds, geckoes, and bats help to control plague conditions of pests, as do ducks, bantams, and a pig to eat waste or fallen fruits. If eel-worms (nematodes) are a problem, plant sunn hemp (*Crotalaria juncea*) and *Tagetes* marigolds throughout the garden beds, one or two every few metres. Sunn hemp root associates trap nematodes, while marigold root exudates suppress weeds and soil fungi, nematodes, and grasses.

5.7
DRYLAND GARDENS

The desert garden is likely to suffer from light saturation and excess evaporation; the former reduces photosynthesis, hence leaf bulk, and the latter causes wilt and slowed growth. To overcome the problems of high pH, heat and light stress, risk of salting in soils, dry winds, and poor water supply, we need to create a special environment around the desert house and garden.

The following are some solutions to the problems of desert gardening:

■ NUTRIENT DEFICIENCY AND ALKALINE SOILS

Plants need three major nutrients to grow well:

1. Nitrogen (N): naturally found in urine, roots and leaves of *Acacia* spp., casuarina, legumes, hair, wool, old woollen clothes or blankets.

2. Phosphorus (P): found in bird and animal manure. Easily collected from under bird roosts and in chicken yards.

3. Potash (K): found in the leaves of comfrey, wood ash, and some volcanic ash.

Plants also need trace elements, and although these may exist in dryland soils, they are usually chemically unavailable to plants due to high soil alkalinity. Mulch and compost are essential to create humus, a soil environment where trace elements can become available. In addition, garden beds should be treated with a light scatter of sulphur to reduce pH to 6.0-7.5. If plants look deficient in trace elements, these can be chemically provided as a foliar (leaf) spray, or added in small amounts to compost rather than put directly into the soil.

■ WIND/SHADE/SUN PROTECTION

Gardens must be carefully sited out of direct windblast, and an extensive use of major and minor windbreaks should be constructed around the house and garden. Wooden fences, tyres stacked 3-6 high, thick-vined trellis structures, and hedges all serve to deflect dry winds. Leguminous trees are acacia, mesquite, albizia, etc., which can be grown on the edges of the garden as windbreak.

To protect young plants from the desert sun, construct a moveable shadehouse out of poles and shadecloth, or plant next to already-existing shade-providing bushes.

Cast light shade over crops in hot deserts by

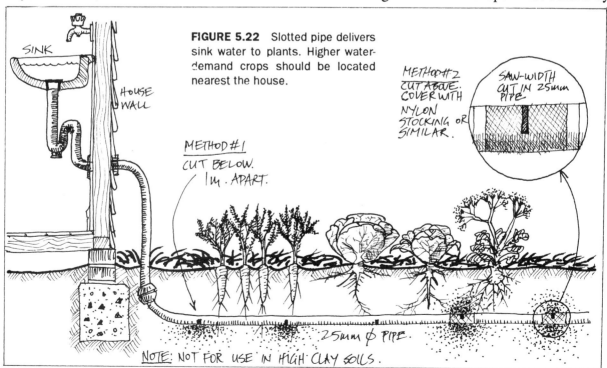

FIGURE 5.22 Slotted pipe delivers sink water to plants. Higher water-demand crops should be located nearest the house.

METHOD #2
CUT ABOVE.
COVER WITH
NYLON
STOCKING OR
SIMILAR.

SAW-WIDTH
CUT IN 25mm
PIPE

SINK

HOUSE
WALL

METHOD #1
CUT BELOW.
1m. APART.

25mm ⌀ PIPE.

NOTE: NOT FOR USE IN HIGH CLAY SOILS.

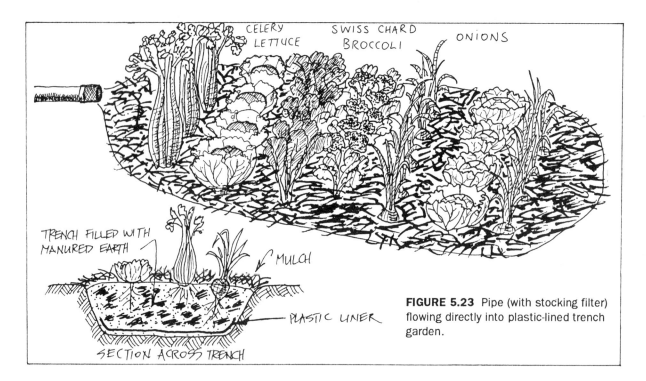

CELERY
LETTUCE
SWISS CHARD
BROCCOLI
ONIONS

TRENCH FILLED WITH
MANURED EARTH

MULCH

PLASTIC LINER

SECTION ACROSS TRENCH

FIGURE 5.23 Pipe (with stocking filter) flowing directly into plastic-lined trench garden.

using spaced vine crop on overhead trellis, or plant open-crowned palms and light-crowned or pruned acacia and mesquite. The trellis system should be integral to the house.

■ WATER

Water is the limiting factor in dryland gardens, but with careful design plenty can be available. Conservation and re-use of water/ wastewater is essential for garden crops, with the handbasin and shower water run into slotted pipes along a plastic-lined shallow planting trench (**Figures 5.22** and **5.23**).

Beds are watered via drip irrigation, preferably below 18cm of mulch or 18cm below the soil surface. Where water is high in salts (most arid areas), it is necessary to apply the water to the surface of flattened mounds or ridges, rather than run it down furrows between crop rows; in the first case, salt gathers harmlessly in the furrows or paths, but in the second (furrows irrigated) it concentrates at the crop roots. **Figure 5.24** shows some garden bed shapes.

Trickle irrigation via commercial pipe systems, or home-made systems of embedded earthenware pots, leaky inverted bottles, gravel-filled pipes are in wide use world-wide. Under tree canopies (citrus, for example), small sprinklers are used in the shaded area to wet 70% or more of the root spread. Sprinklers are, however, not only wasteful on the broadscale but damage foliage by the evaporation of salt on crop foliage, and cause soil crusting. Watering at evening, overnight, or at dawn is preferable to adding water by day due to sun evaporation.

Soil gels can be added at a ratio of 1:100 to garden soils, as can illite (claypan) clays and bentonite clays in sands to assist water retention.

■ MULCH

Mulch is the key strategy for moisture retention and humus build up. Mulch materials are cardboard, newspaper, seagrass, leaves, well-rotted manure, old cotton or wool clothes, sheets of plastic, woodchips, and old carpet or felting. Mulch sources in arid lands may seem bare at times, but in fact there is a great deal of material which can either be grown in the garden (comfrey, legumes), collected after harvest (spent vines and other green material), or gathered from the wild. Trees such as casuarina, pines, and some acacias yield up abundant leaf material. Cattle manures are plentiful at yards and

120

sheds; and near runoff gullies, flood lines leave deep deposits of leaves and twigs. Such mulch is gathered after rains from the creeks and waterflow areas, especially if logs are set at an angle in the creek to trap debris. Stones are often found in drylands, and are useful especially around trees.

Almost any plant does well in the desert garden provided it is adequately watered, which is usually possible only in Zone I and possibly Zone II, as trickle irrigation. Cucurbits, beans, some grains, and both tomatoes and peppers are very successful desert plants for home garden vegetables, as are such desert-adapted trees such as date and doum palm, jujube, mulberry, fig, pomegranate, olive, peach, and apricot. Given good site selection, some basin or swale runoff, and care in establishment, such trees will produce in most seasons over long periods. Thus, an essential long-term strategy is to select adapted plants of low water need, deep-rooted and heat-tolerant.

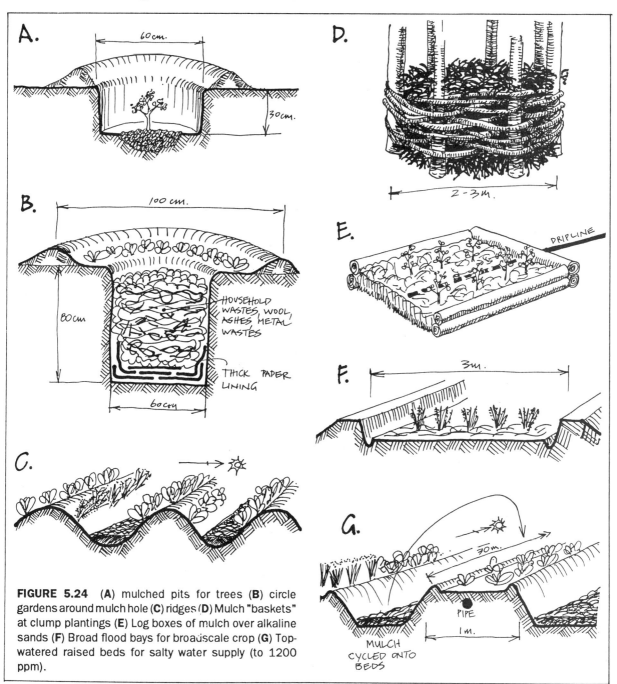

FIGURE 5.24 **(A)** mulched pits for trees **(B)** circle gardens around mulch hole **(C)** ridges **(D)** Mulch "baskets" at clump plantings **(E)** Log boxes of mulch over alkaline sands **(F)** Broad flood bays for broadscale crop **(G)** Top-watered raised beds for salty water supply (to 1200 ppm).

5.8 REFERENCES AND FURTHER READING

Conacher, J., *Pests, Predators & Pesticides (some alternatives to synthetic pesticides)*, Organic Growers Association W.A., 1980

Dean, Ester, *Ester Dean Gardening Book (growing without digging)*, Harper & Row, 1977.

French, Jackie, *Organic Control of Common Weeds*, Aird Books, 1989.

French, Jackie, *The Organic Garden Doctor*, Angus & Robertson, 1988.

Johns, Leslie & Violet Stevenson, *Fruit for the Home and Garden*, Angus & Robertson, 1979. (Out of print; try the library).

Francis, Robyn, *Mandala Gardens Booklet* (with video), 1990, Mandala Gardens, PO Box 185, Lismore Heights, NSW 2480.

Kourik, Robert, *Designing and Maintaining Your Edible Landscape Naturally*, Metamorphic Press, 1986. (PO Box 1841, Santa Rosa, CA 95402, USA).

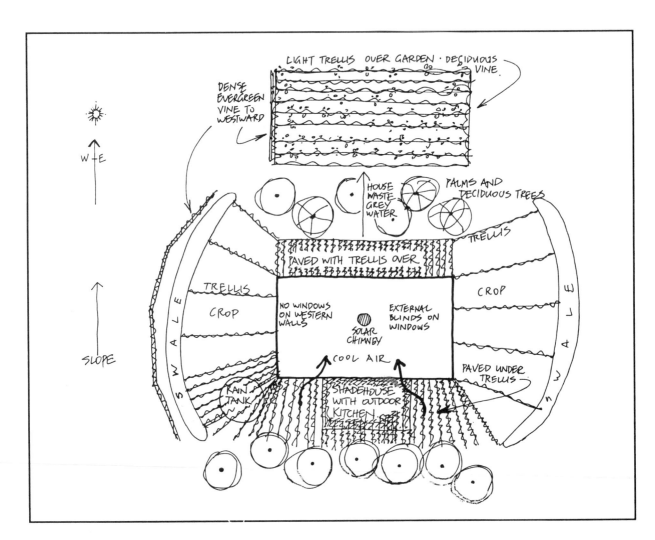

FIGURE 5.25 Idealised layout for a subtropical or dryland garden.

CHAPTER 6

Orchards, Farm Forestry and Grain Crops

Zone II extends out from Zone I and is intensively planned and maintained with spot-mulched or closely-planted orchards, main-crop beds, and ranging domestic animals, whose shelters or sheds may adjoin Zone I. Here we can grow home orchards and grain or main crop vegetables. Commercial orchards and crops are likely to go here and into Zone III, using Zone II mainly for home use. Remember that zones are not fixed, and, in fact, are not strictly delineated. We can put the important elements of a system wherever it is most suits us for easy access.

6.1
ORCHARDS

We can best start the orchard by planting legume (nitrogen-fixing) plants—small species like white clover, lab-lab bean and lucerne, larger species such as acacia, albizia, and black locust, and a scattering of shrubs (tree medic, tagasaste).

Prepare the orchard site by soil conditioning, if necessary, and set out the leguminous species. Interplant selected orchard trees. In home orchards, trees need not be in rows; however, if planning a small commercial orchard, rows are easiest for mowing and harvesting machinery. If planting on a slope, always plant along contours or on contour banks (**Figure 6.1**).

■ PLANNING THE INTERCROP SPECIES

For every element, species and breeds must be chosen to complement the design. Orchards will be made up of disease-resistant main crops (fruit and nut trees), possible windbreak (species that won't compete for light, water, and nutrient), and scattered alternative trees (for pest control, bee attractants). In addition, you will have to decide on the understorey of the orchard. It could be used to grow green manure crops or nitrogen-fixing clovers; provide forage for animals (geese, chickens, sheep); provide a variety of insect- and grass-repelling species; or be used to grow vegetable crop (until eventually shaded out).

Trials of black and red currants, gooseberries, lucerne, feijoa, tagasaste, clover, *Narcissus* spp., perennial dahlia, sunroot, globe artichoke, and the like will reveal successful understorey species for the site. Any deciduous trees removed as diseased can be replaced with evergreen (feijoa, citrus, loquat, olive) and the species mix varied by long-term interplant of chestnut, walnut, almond and plum.

123

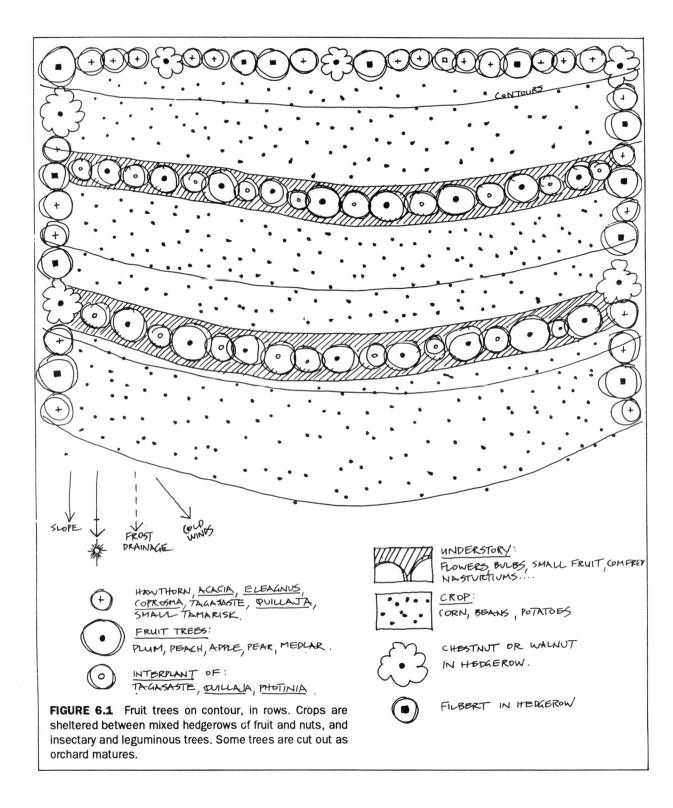

CONTOURS

SLOPE

FROST DRAINAGE

COLD WINDS

⊕ HAWTHORN, ACACIA, ELEAGNUS, COPROSMA, TAGASASTE, QUILLAJA, SMALL TAMARISK.

◉ FRUIT TREES: PLUM, PEACH, APPLE, PEAR, MEDLAR.

◎ INTERPLANT OF: TAGASASTE, QUILLAJA, PHOTINIA.

▨ UNDERSTORY: FLOWERS, BULBS, SMALL FRUIT, COMFREY NASTURTIUMS....

CROP: CORN, BEANS, POTATOES

CHESTNUT OR WALNUT IN HEDGEROW.

FILBERT IN HEDGEROW

FIGURE 6.1 Fruit trees on contour, in rows. Crops are sheltered between mixed hedgerows of fruit and nuts, and insectary and leguminous trees. Some trees are cut out as orchard matures.

Should you be so unfortunate as to inherit a monocultural orchard, add 3-4 hens, a pig, and 4-6 large leguminous trees per 1000 square metres (1/4 acre), with many smaller legumes. For decoration and variety, plant fuchsias, banksias and red hot poker (*Kniphofia*) for the insectivorous birds; borage and white clover for the bees, and add more species as the system evolves. Always try to maximise flowering plants below orchards, as wasp predator refuges.

For a commercial orchard, the same number of fruit trees can be grown, with the area enlarged to fit in the interplant species. Secondary yields such as honey, nuts, foliage, and berries from these added species contribute to the total income. Planned variety gives a good display at wayside stalls and enables direct marketing of varied products, from flowers to fruit to seeds, nuts, and herbs. When deciding which orchard trees are worthy for a commercial venture, select fruit or nuts that:

- bear easily in the climate or microclimate;
- ripen all at once for easy picking;
- ripen evenly;
- have a shelf life and good market value.

When deciding which trees grow best together, it is important to know:

- **The structure of the mature tree**: Is it umbrella-shaped, like mango and walnut, or open like guava and almond? Generally, umbrella-like trees cast dense shade, preventing many crops from growing below them. Open trees, or those with feathery leaves, let enough light to the ground for other crops.
- **Trees that tolerate shady conditions**: Coffee, papaya, hawthorn, most citrus, and black mulberry grow beneath taller trees and may not require full sun to produce fruits.
- **Tree height at maturity**: This is useful to know when deciding a tree's location and space requirement. Smaller trees planted underneath large ones are eventually shaded out, unless pruned severely as is done on the small home garden plots of southern Italy, where mature fig, olive, loquat and even pine trees are pruned to allow sunlight to grape trellis and even vegetable plots (growing between grapes).
- **Moisture needs**: Place drought-resistant trees (carob, almond, guava) and moisture-needy plants (papaya, banana) in separate groupings to aid in watering.
- **Allelopathy**: Make sure selected trees do well together. Walnuts, for example, secrete a substance in their roots which cause many fruiting trees to grow poorly.

We should also consider the need for cross-pollination, placing male and female species of the same tree near each other.

■ ANIMALS IN THE ORCHARD

Once young orchard trees and their associated plant guild species are established, small livestock can be introduced. In the beginning, bantams and small poultry breeds can be on range. Poultry scavenge most soft fruits (and any larvae or pupae of pests), help control weedy vegetation, provide manures for orchard trees, and forage for seed and greens. Chickens at 120-240/ha do not greatly affect the density of shrubby ground covers. When orchard trees are 3-7 years old, foraging pigs can be introduced as fruit matures to take care of windfall fruits that breed pests. In standard-pruned orchards of from 7-20 years old, first sheep and later controlled cattle grazing can be permitted. Watch to see that sheep and cows do not damage tree bark; if so, they must be removed or the trees protected.

■ TEMPERATE PLANT GUILD FOR THE FRUIT ORCHARD

The enemy of deciduous orchards is grass, thus non-grass crop below tree canopies is ideal (**Figure 6.2**). A mix of the following plant groups can be made:

- **Spring Bulbs** (daffodils, hyacinth): These flower and die back by early summer, as do most of the onion species (*Allium*), and create a grass-free area below trees in fruit, plus a crop of bulbs, flowers, and honey. Iris and tuberous-rooted flowers also assist grass control.
- **Spike Roots** (comfrey, dandelion, globe artichoke) cover the ground and encourage worms, yield mulch and crop. Soil below their foliage is soft, free-draining, open to roots feeding near the surface, cool.
- **Insectary Plants** and small-flowered plants: fennel, dill, Queen Anne's lace, tansy, carrot, and parsnip flowers (umbelliferous). Predatory wasps, robber flies, ladybirds, jewel beetles, and pollinator bees are attracted to interplants in the orchard. In the herb layer, catnip, fennel, dill, small varieties of daisy (or any of the Compositae family), and flowering ground covers generally attract wasps, bees and insectivorous birds.
- **Nitrogen and Nutrient Crop**: Clovers and

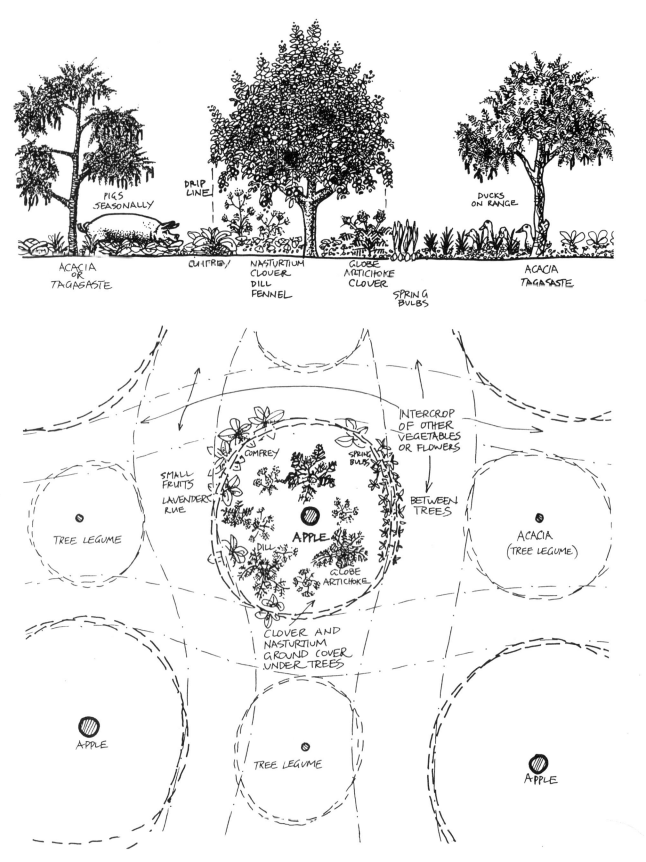

FIGURE 6.2 Idealised guild assembly for an apple orchard. Legume trees are lopped for mulch; perennial and annual flowers help in pest control; grass is eliminated by comfrey and herbs.

interplants of tagasaste or acacias provide root-level nitrogen. Marigolds (only the *Tagetes* species) planted around trees "fumigate" the soil, as does the green crop sunn hemp for nematodes.

Such guilds of plants are needed especially in the first few years of orchard establishment. Trees of 10-plus years of age are far less susceptible to grass competition, and thus groundcover guild plants are less needed.

In general, we aim to reduce or even eliminate grasses, to plant as many flowering plants as possible to attract a variety of pollinators, predatory insects, and insectivorous birds (using *Kniphofia* spp., *Fuchsia* spp., *Echium fastuosum*, *Salvia* spp.) and to provide ground cover, stone piles, logs, pits, and tussocks for frogs and insectivorous lizards. Small ponds throughout orchards will breed frogs for leaf insect control.

Soft ground covers such as nasturtium prevent the soil from drying out and give mulch, as do the interplant and windbreak trees, and the herb layer generally.

To sum up, pest species in the orchard can be reduced by a combination of these strategies:

• Selecting disease resistant stock for the main fruit crop;

• Planting flowering crops and refuges for predation by birds, frogs, lizards, wasp and predatory insects;

• Interplanting leguminous trees and small trees other than the main crop species;

• Reducing orchard stress by removing grass cover and protecting with windbreak and mulch; and

• Ground foraging by chickens, pigs, geese to clean up windfalls and deposit manures, or the careful collection of windfalls for juice processing or disposal.

■ TROPICAL ORCHARDS

A mix of tree legumes, fruits, bananas, papayas, arrowroot (*Canna*), cassava, sweet potato, and comfrey can be co-planted on loosened soils and in mulched swales. There should be large species planted every 8-10 metres (mango, avocado, jakfruit) with smaller species (citrus, tamarillo, guava) interplanted with coconut during the establishment period. Smaller shrubs and plants are planted as gap fillers (**Figure 6.3**).

The planting area around the small trees can

PRODUCTION PALMS

THIN-FOLIAGED LEGUMES

FRUIT NUTS

FRUIT NUTS

COFFEE
PINEAPPLE
GINGER
CASSAVA
SWEET POTATO, ETC

FIGURE 6.3 The home orchard layout can be similar to tropical "stacking" of plants found in rainforest, where plants of varying heights share light and nutrients. In this type of system, a supply of water from dams is necessary to get through the dry season (if rain does not fall all year around).

also be seeded with nasturtium, lab lab bean, Haifa clover, broad bean (*Fava*), buckwheat, dill, fennel, lupin, dun peas, pigeon pea, or any useful non-grass mix available, and suitable to climate, landscape, and available water. The aim is to completely carpet and overshade the ground in the first 18-20 months of growth.

Ideally, dense plantings of this type should be sheet-mulched with newspapers/cardboard and topped with available cut grasses and later the tops of arrowroot, comfrey, banana, acacia, and green crop. Later still, shade-loving species such as coffee and dry taro can be placed in any open spots. Turmeric, taro, ginger, sweet potato, and cassava are cropped below tree systems.

It is far better to occupy a quarter hectare thoroughly than to scatter trees and herbs over a large area. Much of the small vegetation is used for mulch and nutrient, and should be thickly applied to suppress grasses.

When planting on slopes, trees should be on the contour, with strip plantings of *Canna*, vetiver grass, lemongrass, or elephant grass. These are set out to form an unbroken cross-slope hedge, or across on earth walls or dam banks at spill-ways. They disperse water and create silt traps; behind such self-perpetuating walls, soil is deeper and trees can be planted.

When pioneering in grassland or expanding the systems outwards, use wet hollows, small dams, and cross-sloped swales to hold wet-season water (**Figure 6.4**). Around these plant hardy legumes such as leucaena, *Inga, Acacia mearnsii* and other acacias, Gliricidia, Calliandra, *Cassia, Gmelina,* Albizia, *Bauhinia,* Tamarind, etc. All of these withstand grasses after the second year.

Weed species such as lantana and *Pennisetum* give excellent early cover, and are later cut to make rough mulch piles 3-6 metres across in which vines, palms, and useful legumes are far more easily established. Also use rampant and vigorous soft vines (choko, yams, passionfruit) to clamber over and shade out shrubby weeds, which are later slashed for use as tree mulch.

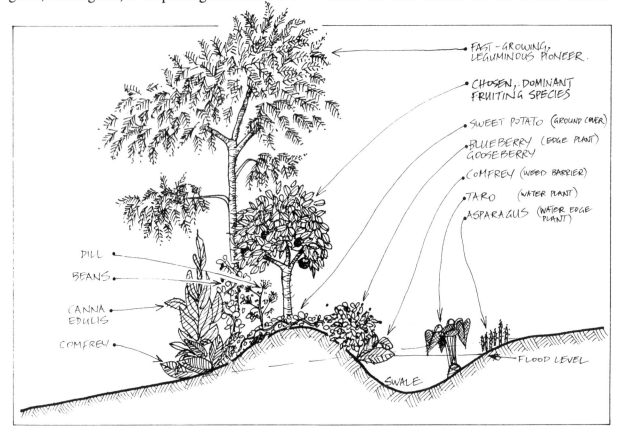

FIGURE 6.4 Trees planted off swale bank to take advantage of wet-season water.

SAMPLE SPECIES LISTS FOR SUBTROPICAL ORCHARD INTERPLANTS

Setting up the orchard guild means that first ponds, swales, access roads, tracks, large circular mulch pits, and sub-guild areas or natural openings are developed and connected to infiltrate water and to provide good wet-season drainage (mounds for avocado and citrus, banks and small mounds for yam and pineapple). Waterlines and taps are set in for the essential two to three years watering of the young trees during the dry season. It is far easier to place these permanent elements at the beginning than to work around the plants once they are in place.

The sub-guild areas are best kept small (300-400 square metres) and "edged" with short perennial species such as lemongrass and comfrey (cut seasonally for mulch) or with swales. Larger plants (pioneers, large trees, windbreak legumes) can be planted throughout, widely-spaced, and later each bloc is fully-planted and mulched, one small bloc at a time.

Major windbreaks: Silky oak, casuarina, pongamia, sesbania, *Prosopis* spp., Golden Goddess bamboo (a clumping variety), makes a dense windbreak when planted in a mulched line or arc.

In-crop windbreaks: Gliricidia, *Tipuana tipu*, and acacia planted as individual trees which may eventually be cut out and used as mulch.

Pioneer plants: Usually already growing on site and useful as shade, mulch, and soil improvement, e.g. lantana; wild tobacco bush, macaranga, acacias.

Legumes: *Albizia* spp. and *Acacia* spp. (*A. fimbriata*, *A. auriculiformis*, *A. longifolia*), leucaena, *Inga edulis* (ice cream bean tree), sunn hemp, *Cassia multijuga* and spp.

Large trees: Macadamia, mango, jakfruit, pecan, avocado, lychee.

Medium trees: White and black sapotes, custard apple, olive, fig, rambutan, loquat, carambola, mulberry, persimmon.

Smaller trees: Tamarillo, citrus, feijoa, coffee, papaya, banana, jaboticaba, small guavas, rosella, rose apple, Brazilian cherry.

Palms: Date palm, coconut, butia, adapted tall palms generally.

Trellised vines: These can be placed between Zone I and II, and in and around the orchard in the first few years. Black and yellow passionfruits (5-6 good varieties, including lillikoi), choko (chayote), grape (species suited to subtropics are available), oysternut (a vigorous cucurbit yielding a nut, grown extensively in southern Africa), kiwifruit, luffa, a variety of beans, and gourds.

Tubers and ground covers: sweet potato (permanent groundcover; may be harvested every few years if necessary), turmeric, ginger, cardamon, *Canna edulis* (Queensland arrowroot), *Chilecayote* (a rambling perennial squash which must be cut off trees occasionally), pigeon pea (*Cajanus cajan*), pineapple, comfrey, lemongrass.

Wet and swampy areas: Chinese water chestnut, taro, *Sagittaria*, lotus, waterlily, tea tree. Bananas do best near a greywater outfall.

In the tropics and subtropics, nutrients are cycled through the vegetation, not the soil, hence the emphasis on mass plantings and stacking of vegetation layers. If the plantings get too dense (as the larger trees gain maturity), simply chop, mulch some of them out, or transfer them elsewhere. The orchard, especially in its first five years, is a dynamic and changing component of the system and the components of lower storeys can be split up by bulbs, cuttings, divisions, and so on.

Water demands are greater in the first few years; however, almost all the species mentioned above are dormant or slow-growing in winter, the dry season in the subtropics. Watering may be necessary in the few months before the summer rains, although a fully-mulched and shaded orchard will not need as much water as one with bare ground.

After 2 to 3 years of leguminous tree culture, there will be a great improvement in soils; after 3-7 years, a high thin canopy of palms, feathery legumes, or wet-season deciduous legumes (e.g. *Acacia albida*) will enable a complex assembly of understorey vines, shrubs, trees, and strip crops to flourish.

Pigeon pea, cow-pea, daikon radish, clover, and lucerne can be broadcast and raked into areas of disturbed soil around tree seedings. These all loosen and create humic (organic) soils.

Species which grow from large cuttings (some mulberries, horse radish tree, any local species) can be placed at the edge of clumped forest plantings, as these can be quickly propagated after a few years by coppicing.

■ DRYLAND ORCHARDS

Any dryland area will support fruit and nut trees if there is an adequate supply of water. Trees set out in drylands include date palm, jujube, cork oak, pistachio, plum, white cedar, tamarisk, chestnut, honey locust, carob, tagasaste, mesquite, paulownia, with bulk cuttings of grape, fig, and mulberry. Other plants which withstand dry conditions are apricot, almond, pomegranate, olive, and cactus (*Opuntia*

spp.) These represent a range of fruit, nuts, nitrogen-fixing legumes, and other useful species (**Figure 6.5**).

Because of the lack of water in drylands, plants are not crowded together as in the tropics, and in fact orchards usually mimic the plant structure of natural drylands, where plants are spaced so they do not compete for water and nutrient. All important trees should be mulched and put on drip irrigation.

In stony deserts or dry slope areas, where surface stone is readily available, stones alone make a permanent mulch around trees. In the Canary Islands, volcanic pumice is spread liberally in orchards as a stone mulch. Stones are of benefit to plants by:
• protecting and shading roots from intense day heat;
• releasing stored heat to the soil at night;
• preventing poultry or small animal damage to roots;
• preventing wind lifting of roots;
• providing shelter for worms and small soil organisms;
• causing water to condense on their surfaces on very cool nights.

The most successful dryland tree-planting

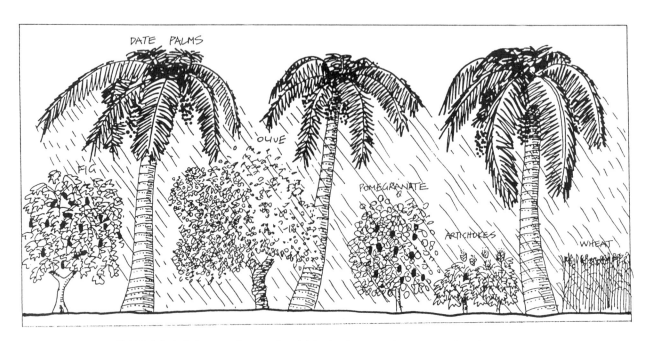

FIGURE 6.5 Palms can be used as high shade over other tree and surface crops.

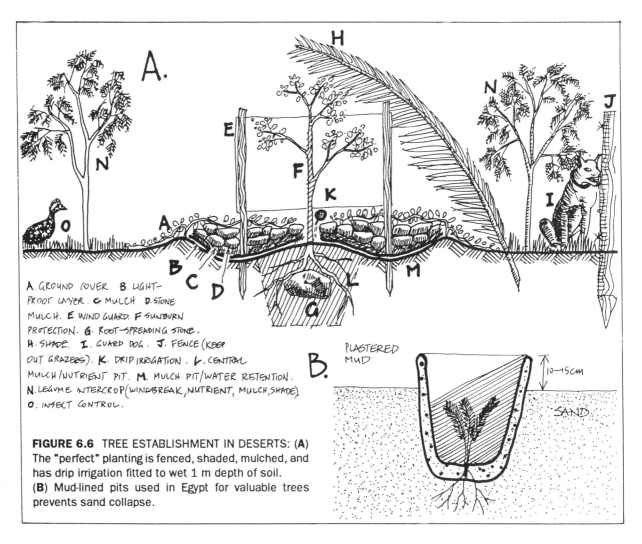

A. GROUND COVER. B. LIGHT-PROOF LAYER. C MULCH D. STONE MULCH. E WIND GUARD. F SUNBURN PROTECTION. G. ROOT-SPREADING STONE. H. SHADE. I. GUARD DOG. J. FENCE (KEEP OUT GRAZERS). K. DRIP IRRIGATION. L. CENTRAL MULCH/NUTRIENT PIT. M. MULCH PIT/WATER RETENTION. N. LEGUME INTERCROP (WINDBREAK, NUTRIENT, MULCH, SHADE) O. INSECT CONTROL.

PLASTERED MUD

10–15 CM

SAND

FIGURE 6.6 TREE ESTABLISHMENT IN DESERTS: **(A)** The "perfect" planting is fenced, shaded, mulched, and has drip irrigation fitted to wet 1 m depth of soil. **(B)** Mud-lined pits used in Egypt for valuable trees prevents sand collapse.

strategy is to plant on the edges of swales. House-roof water and stormwater drains lead into the swales, which will slowly filter down into the earth. Road runoff and streamflow can be led into tree-lined swales to great advantage.

The following is a checklist for planting valuable trees in drylands:
• Select suitable tree species for the area; if native, give preference to local seed.
• Plant well-grown trees for a better survival rate.
• Plant in the rainy season to ensure enough water getting to the tree.
• Plant trees and shrubs together in a group planting, yet not so close that they will compete when growing.
• Install a drip irrigation system to each tree. Water deeply and slowly to encourage roots to go deep into the ground to find their own water.
• Hold water around the tree by making a shallow basin, lining it with newspaper and placing straw and then rocks on top for the slow release of moisture.
• Suppress all grass from around the tree by mulching; other suitable small plants can grow in the mulch.
• Protect the tree from sunburn, windburn, and animals by shading, planting windbreaks or bagging, and fencing (**Figure 6.6**).

Planting on Hills

The "net and pan" planting pattern of **Figure 6.7** is an effective erosion control in overgrazed, eroded, mined or bulldozed sites. If tyres are available, the "pans" can be made from these, filled with mulch, and the diversion drains led in above the tread level. If logs are available, these

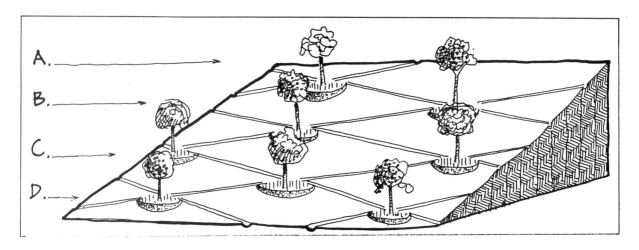

FIGURE 6.7 NET AND PAN PLANTINGS FOR ARID HILLSLOPES. **(A)** Crest trees: hardy needle-leaf species and narrow-leaf trees to suit thin soils, e.g. stone pine, olive, casuarina, acacia **(B)** Hardy trees with known drought resistance, e.g. fig, Spanish chestnut, pomegranate, acacia **(C)** Mid-slope and deeper soils suited to citrus, fig, pistachio **(D)** Deeper soils with some humus, suited to mulberry, citrus.

are staked cross-slope, on a slight downhill grade so that water is made to zig-zag across the erosion face, and hence absorb into the ground. Even small logs and branches, pegged across erosion channels, build up a layer-cake of silt and leaves, beside which tagasaste, acacia, or any other fibrous-rooted and hardy species can be planted, which then act as a permanent silt trap. Mulch behind logs and barriers quickly stabilises the area for planting.

On very steep slopes there is often no recourse other than to plant bamboo and rootmat pioneers, and to make upslope plantings of chestnut, acacia, carob, olive or other large species which will cascade seed downslope over time.

Corridor Planting

Although home orchards should be close to the house and to a water source, another method of establishing tree systems in drylands is to depart from strict zone and sector design in Zone II to IV, and adopt a more flexible strategy of *corridor* development. This follows valleys and steam beds, and intermittent stream flow to take advantage of shade, water, and mulch accumulation. Thus, from Zone II outwards, we plant our trees along the corridors of flows developed by water systems, planting hardy trees along the edges of river beds and in the shady valley flows of wadis. Palms and dates, in particular, like the sandy edges of stream beds.

By observing the way plants already grow in nature, we can place new plants with a great deal more success than by trying to bring water to dry land and use it there. Areas of bare rock act as runoff surfaces to concentrate water in soils, and by finding these naturally-damp or rich-nutrient sites, we can grow almond, olive, citrus, chestnut, bamboo, mulberry, fig, and date in suitable areas. This gives us a lot less work than laying out a "forest" on a flat site, as the trees are grown where they are suited.

There are several broadscale methods of establishing trees in drylands. Tree seedlings from the nursery are set out, during the rainy season, with minimal preparation and follow-through (except for stone mulch) to see which species will grow, with neglect, in arid regions. This strategy will work best if the area is fenced to keep out hoofed animals and other browsers.

Another approach is to mud-pellet seed, using an old meat mincer with the blades off, and a damp slurry of mud, rock phosphate, urea, and seeds. These are put through the mincer and rolled or shaken in dry dust to form pellets, which are then carefully buried in likely sites for trees, to await rain. The mud keeps birds and ants from eating the seed pellet.

There are many ways of placing trees in arid areas where they will get a chance to grow. In rock gullies, around rock domes, in sandy runnels between rocks, on dry, stony hillsides, on the banks of creekbeds and floodplains, trees can be planted with success.

The object of semi-arid sustainable broadscale design is to achieve these ends:

1. To exclude large browsers from crop, crop-savannah, or orchard by planting thorny and inedible hedgerow.

2. To diversify hedges with large useful timber, forage trees, and low useful flowering shrubs to harbour birds and predatory insects, and to provide a place for cucurbits and other yielding vines, beans, and fruits.

3. To exclude drying winds by main windbreak of 5-8 trees wide, every 50-100 metres; an in-crop windbreak of legumes every 30 metres or so, and tall strip crop interplant to act as smaller in-crop windbreak every 2-10 metres as either alley cropping at 2 metres or broad strip crop at 5-10 metres.

4. To shape the ground to harvest all overland flow from rains and to absorb it into the ground; all such systems need to be able to hold 10-30cm of rain in one continuous episode, and to soak it in from 2 to 40 hours. This can be done through swales, pitting the ground, walled fields on floodways, and terraces on low cropped slopes.

6.2
STRUCTURAL FORESTS

In recent decades farmers have started developing tree systems on farms, which have seen a change from annual cropping to a mixed crop of annual grasses (or crops) and trees. There are many reasons for this change, including:

• The realisation that trees provide forages in hard times for both livestock and wildlife, and that they buffer conditions of extreme heat and cold.

• Concern about soil erosion on steep slopes and along watercourses. Trees also lower the groundwater table and so prevent salting of soils.

• The need to diversify farm products, which buffer economic changes in prices for crops and livestock. Early diversification can be into honey and pollen production, with later diversification into a wide range of animal and plant products (fruit, nut and vine products).

• The need to have a source of on-farm firewood and building materials.

• Concern for wildlife refuge areas, especially for birds, important to crop pest control.

Layouts of farm forests can vary depending on machinery or labour available, landscape features, and farm priorities or purposes. Some systems are as follows:

■ TIMBER CROP IN PASTURE

Selected high-value trees are widely spaced in rows to permit good pasture development between the trees. Ideally, tree rows are on contour. Animals are let in to browse pasture when the trees are hardy (timing usually depends on species planted), and before that grasses are harvested for hay or silage. Crops may even be grown or the area cover-cropped continuously to build fertility.

Some successful pasture-timber intercrops include black walnut, some pines (*Pinus pinaster*, *P. caribaea*, *P. elliottii*), poplar, paulownia, silky oak (*Grevillea robusta*), white cypress pine (*Callitris columellaris*). Some of these may need management (lopping lower limbs, etc.) to be worthwhile timber.

■ FIREWOOD PRODUCTION

Firewood comes as cones from nut pines, fallen wood, coppice, forest thinnings, or pioneer trees cut out at the end of their useful life. As the forest matures, however, these types of woods become less available, and the system should be expanded by frequent replanting for permanent yield.

Woodlots are often planted on farms to pro-

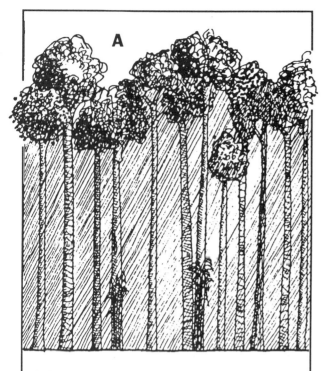

A. Dense forest stand: Maximum number of trees per unit area. Straight stems; first-class timber. Full, closed canopy; little understorey possible. Minimal bearing surface.

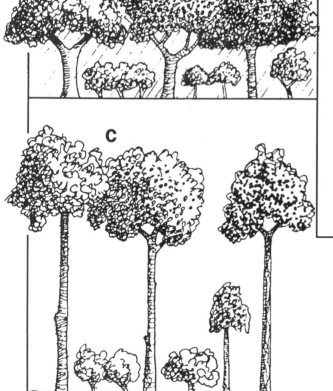

vide a fast turnover of firewood. These are usually on a 2-7 year rotation (one-half to one-seventh of the trees cut annually). Depending on the tree, firewood can be cut as coppice or stickwood, or grown to 4-10cm log size. In most cases, firewood species are chosen for persistent coppice ability (regrowth from stump) and good fuel value. Some eucalyptus and acacia species have this ability.

■ POLEWOOD

Polewood is important for fencing, and house and furniture construction. Durable woods for outdoor uses are chestnut, raspberry jam acacia (*Acacia acuminata*), osage orange, black or honey locust, cedars generally, and eucalypts known to be rot resistant (river red gum, turpentine). Polewood of less durable quality is used for indoor work, furniture, and as scaffolding or formwork support in building.

■ LONGTERM FINE TIMBERS

A section of the farm can be reserved for growing longterm timbers such as black walnut, rosewood, teak, cedar, blackwood, oak, redwood and any local fine timber. Although these can be planted on land not readily useful to the farmer, they need to be managed to maintain straight trunks. **Figure 6.8** shows different forest stands due to different plant spacing, species, and management.

Some very valuable trees, such as black walnut, not only produce young trees as poles, but may be sold out as rootstock for grafting, and at maturity enable the farmer to retire on the sale income. Timber trees can be interplanted with fast-growing, multi-use species. Black locust, for example, is a pioneer tree and good soil

FIGURE 6.8 MANAGEMENT STRATEGIES FOR FOREST STANDS
B. Open-growth forest: Minimum number of trees per unit area. Dense canopy, but some understorey possible. Large bearing surface per tree. Poor timber quality.
C. Thinned forest of type (**A**) above: Yields pole timber as thinnings. Remaining trees yield good timber. Open canopy permits understorey. Maxium bearing surface per tree.

builder. A durable wood, it will grow to post-height in 6-10 years, and can also be coppiced for fuelwood. Lastly, it provides chicken forage.

Bamboo is another wood that has numerous domestic uses. Although slow-growing, large clumps can be broken up and propagated for faster production. Species of bamboo grow from the temperate regions to the tropics, with tropical and subtropical species large enough to be used as scaffolding, furniture, gutters, and reinforcement for concrete. Bamboo shoots are also eaten, and the small leaves used as mulch in the garden. Care must be taken, however, in situations where bamboo will spread and displace important native vegetation, particularly along watercourses. It is best to use a clumping bamboo rather than a running type.

■ HEDGEROWS

Shelterbelt, hedgerow, and animal barrier forests systems have special shapes as windbreaks around the house and farm site, and shelter for animals against heat and cold. Hedgerow and windbreak species are chosen for fruit and nut yields, forages, honey, special wildlife foods, browse, and both mulch and stickwood production. Unlike some other forest types, hedgerow and windbreak forests can contain numerous species, as the trees are not cut down for their product, but rather fruits and nuts are selected and gathered. Barrier hedges made of thorny, unpalatable, or impenetrable plants keep most livestock from straying into gardens and field crops. See Chapter 2 for a full commentary on windbreaks and hedgerows.

To build up a mixed forest, the essential precursors are the pioneer species. These are the fast-growing, leguminous trees that build the soil, and provide mulch and shelter for slower-growing trees. And depending on the species selected, they also provide nectar for bees and seed for poultry forage, with their branches coppiced for firewood.

Trees are established in clumps (fed by several drip-points if necessary), as this allows them to self-shelter and spread by seed. Indi-vidual plantings tend to get ignored, and often dry out, suffer windpruning, and are smothered by grass competition.

Understorey shrubs are an important part of the forest system, as they help to establish microclimatic conditions and aid in grass suppression. Leguminous shrubs enrich the soil, and are necessary in any cut-and-take system. All forestry should be designed as a multi-tiered canopy and plants chosen to yield many products. Forest products other than wood are mulch, mushrooms (shiitake), honey, herbal medicines, and oils.

■ NATURAL FOREST

In any forest we should leave a section that is not managed; it is left in its natural state for wildlife habitat and forage, and to protect fragile upper slopes against erosion. These undisturbed areas are very beautiful, peaceful places, and of intrinsic worth. We are able to contemplate nature here, and to learn about ourselves in the natural world. Those of you who have been alone in a forest for a long time—more than five weeks—know that you can totally lose your identity as a human being. You can't distinguish yourself from the trees, from the animals, or from any other living thing there. All aboriginal people, tribal people, have to undergo such a period on their own in the environment. Afterwards, they never again perceive themselves as separate: me here and tree there. You become simply a part of all life.

Tropical forests are of great diversity and of great importance in the health and maintenance of the global atmosphere. A grave error is to settle permanently in such a forest and clear any part of it (as is now being done in Brazil and Sumatra). Far better to make the already-settled areas more productive, and to govern population increase.

Protection and enlargement of remaining forests are not only a global but an individual concern. Forest is the greatest resource on earth; value it for its many gifts of medicine, clear water, breathable air, and materials for our future, and its honey, diversity of species, rubber, and nuts that can be gathered only from living trees.

The following sections contain examples of a temperate and a tropical grain-growing system. These can be as small or as large as we like, and placed in Zones II or III according to size and access.

■ FUKUOKA-STYLE GRAIN CROPS FOR TEMPERATE REGIONS

Until I read Masanobu Fukuoka's *The One Straw Revolution*, I felt there was no satisfactory basis for including grain and legume main cropping in permaculture. However, this system has solved the problems of no-dig grain culture.

In brief, the system combines the usual rotation of legume/grain/root crop/pasture/fallow/legume into a single grain/legume mixed crop. The idea is to broadcast the next crop *into the maturing crop*. The system uses the principle of continuous mulch (with clover) plus double-cropping using winter and spring sown grains. This is what makes it possible to use small areas (400 square metres or less) to supply a family's grain needs.

If paddy rice is to be grown, the area must first be graded or levelled, and a low bund (water-retaining wall) built around the plot, so that 5cm or so of water can lie on the ground in the summer.

After levelling or preparation, lime or dolomite is spread over it, watered in, and made ready for autumn planting (**Figure 6.9**). I will deal with more than one plot here, to show how different plants can be treated.

In autumn, seed is broadcast as follows:
Plot 1: Rice, white clover, rye.
Plot 2: Rice, white clover, barley
Plot 3: Rice, white clover, millet
Plot 4: Rice, white clover, winter wheat
Plot 5: Rice, white clover, oats

The rice lies until spring, and other crops germinate soon after sowing.

Early autumn: A thin layer of chicken manure is broadcast over the area. Use clover at 1 kg per ha (1 lb/acre), rye and other grains at 7-16 kg per ha, and rice at 6-11 kg per ha (5-10 lbs/acre). Use inoculated clover if this is the first crop. The seed can be scattered first, then straw-covered to protect from birds. Alternatively, the grain seeds are mixed with mud, pressed through wire-mesh and rolled into small balls, or dampened and shaken in a tray of clay dust to form mud-coated pellets. In the second year, rye and clover are sown into the ripe rice crop at this time.

Mid-autumn: Last year's rice is reaped, the crop dried on racks for 2-3 weeks, and threshed. All rice straw and husks are returned to the field. Unhusked rice is now resown within a month of harvesting, just before the straw is returned.

Winter: Light grazing of the winter crops by ducks assist the stooling of plants and will add manure. Check and sow any "thin" areas as soon as possible. When the crop has reached 15cm or thereabouts, about 100 ducks per ha (40/acre) will reduce pests and add manure. Fields (or paddies) are kept well-drained during this time.

Spring: Check that rice is growing, and re-sow thin patches if necessary.

Late spring: Rye, barley, etc. is harvested and stacked to dry for 7-10 days. The rice is trodden, but recovers. When other grains are threshed, all straw and husks are returned to the fields, moving each straw type on to a different plot thus:

Plot 1: Oats   Plot 4: Millet
Plot 2: Rye   Plot 5: Wheat
Plot 3: Barley

Early summer: Only rice remains. Summer weeds may sprout; these are weakened by flooding for 7-10 days, until the clover is yellow but not dead. Rice grows on until harvest.

Summer: The field is kept at 50-80% saturation under rice, while seeds of other grains are prepared for sowing in early autumn. The cycle then continues as before, but now using the crop straw for mulch.

136

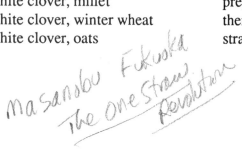
Masanobu Fukuoka
The One Straw Revolution

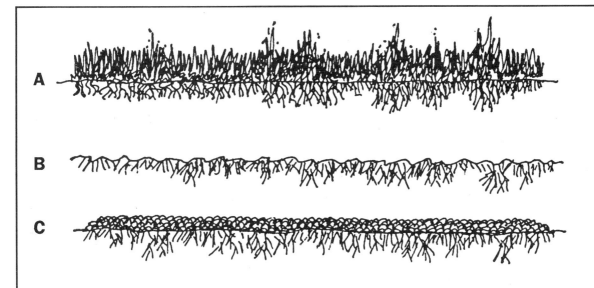

A

B

C

1. AUTUMN: Original surface is mown (**A**), chisel ploughed (**B**), and mulched (**C**). Essential manures are added, and rice, rye, and white clover broadcast sown.

2. WINTER: Clover and rye sprout and grow, rice lies dormant in husks.

→ HARVESTED.

3. SPRING/SUMMER: Rye ripens and is harvested. Rice sprouts and grows. Rye straw is returned to field. Later rice is kept saturated and is harvested. Rice straw is returned.

4. AUTUMN: Commence cycle again as at **1C**; vary crop to millet, wheat, beans, lentils, etc.

FIGURE 6.9 Schematic of no-tillage grain and legume cropping.

Each person must evolve their own techniques and species mixtures, but once a cycle is perfected there is no further cultivation, and straw mulch is the only weed control. It helps if the area of bunds around the crop is planted to *Coprosma*, comfrey, citrus, mulberry, lemongrass, tagasaste, pampas, or other weed-controlling shelter plant. Mulch with sawdust under these borders to prevent weed re-invasion from the bunds or surrounding land.

Where a paddy is not possible, dry-land rice or other grain species can be used, and spray irrigation replaces summer flooding. In monsoon areas, summer rain should suffice. Where rice cannot be grown (e.g. very cool areas) other grains may be substituted and short-term cycles may be developed (spring wheat or corn sown in early spring, for example, with oats, barley or

137

wheat as winter crop). Other legumes can also be tried.

Further useful systems and data is given in *No-Tillage Farming* by Phillips and Young, Reiman Associates, Wisconsin, 1973 (a book unfortunately oriented to heavy machinery and chemical sprays). Rye and wheat are broadcast into soybean crops when the leaves on the latter begin to fall—the falling leaves hide the seed from birds. Soybeans (or other legumes) are broadcast into the stubble of oats, barley, wheat, or rye, as is lespedeza, which is autumn-harvested. Peas are planted after corn, and green peas are followed by corn. Other crops suited to no-tillage are cucumber, watermelon, tomato, cotton, tobacco, sugar beet, capsicum, vetch, and sunflower.

Fukuoka's book gives much more data on no-tillage gardening for vegetables and fruit.

For the tree crops he used 12 acacia trees (silver wattle, for example) to the hectare (5/acre) instead of clover. He has maintained this no-dig cycle for 35 years, and his soil has improved with no fertiliser other than chicken and duck manure, no sprays, and no herbicides.

■ AVENUE CROPPING TECHNIQUES FOR MONSOON TROPICS

Avenue cropping is the growing of crops between rows of frequently-pruned legume trees such as leucaena and gliricidia, and using the branches and leaves of these to provide fertiliser and mulch for crops. As the mulch layers decompose, they contribute valuable nutrients to the soil and feed earthworms.

A main crop area of rice, mustard seed, taro, wheat, maize, potatoes, etc. can be grown in 2-4 metre strips between the legume hedges, which

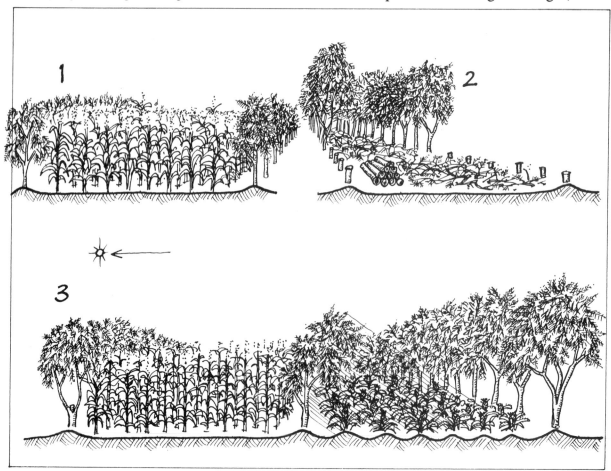

FIGURE 6.10 AVENUE CROPPING: (**1**) Both crop and tree legumes are planted (**2**) Branches are lopped and used as mulch in the crop; when the crop is harvested, trees may be lopped to the ground for firewood (**3**) Cycle is repeated; trees may shade crop, which is desirable in some situations.

are repeatedly cut back to 0.3 metres to sprout again. Winter (cool dry) crops are mustard, wheat, clover mulches, millets; wet season crops are maize, rice, taro, beans. Semi-commercial crop is ginger, turmeric, pineapple, melons, and gourds. To reduce risk of soil diseases, plan to rotate bed use so that, for example, potatoes move every year over a 5-year period.

The ground is prepared by digging and is laid out in long mounds along the contour. Ideally, soil deficiencies are corrected at this time, with some blood and bone added, and the area is mulched with straw. Both crop and leguminous trees are planted, as shown in **Figure 6.10**.

At the International Institute of Tropical Agriculture (IITA) in Nigeria, studies show that *Leucaena leucocephala* and *Gliricidia sepium* can be cut five times a year for seven years before they have to be replaced. Depending on needs, these hedges are allowed to grow on (leaving the crop area fallow or planting to shade-tolerant species such as pineapple) to produce livestock fodder during the dry season. Grown with such grasses as *Panicum maximum* and *Pennisetum purpureum*, they supplement the feed of sheep and goats on a "cut and feed

out" basis.

Some rows can be left to grow to sapling size for stick firewood, useful in countries where firewood is needed for cooking.

The idea of avenue cropping should be not limited to the tropics (although it is most suited for that climate because moisture and warmth make for increased vigour). Cut-and-mulch or cut-and-feed systems have been developed for temperate climates and include tagasaste, poplar, and willow.

■ TRADITIONAL INTER-CROPPING SYSTEMS OF A DRY MONSOON AREA

The Deccan is an arid area in southern India where many small-scale farmers plant field crops in traditional ways using non-hybrid seeds. The traditional small fields of the Deccan form a crop guild with its accompanying tree-crop trees and hedgerows which supply honey, nitrogen (legumes) and fruit and nuts, and consist of the following broad groups growing together:

• **Main crop**: usually a grain, grain legume, or tuber/root crop such as sorghum, millet, maize, rice, wheat, oats, barley, rye, potato, cassava, sweet potato, turmeric, ginger, chick pea, pi-

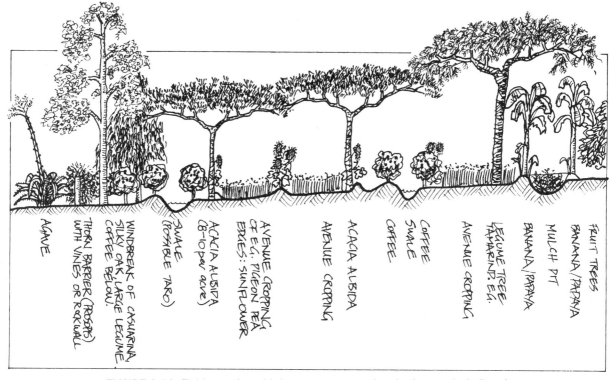

FIGURE 6.11 Field cropping with legume trees, swales, hedges and windbreak.

139

geon pea, black gram, horse gram, mung bean.

• **Legumes**: Trees, shrubs or vines providing nitrogen and humus to soils, micronutrients from leaves, honey, and predator refuge. Trees are *Prosopis* spp. *Acacia* spp. sesbania spp., *Cassia* spp. gliricidia, pongamia. Smaller legumes are velvet bean, cow pea, pigeon pea, vetches, clovers, winged bean. Permanent trees planted at 35-50 trees per hectare stand in the crop fields all the year.

• **Flowers**: Often herbs of the family Umbelliferae (dill, fennel, coriander, etc.) and Compositae (sunflower, marigold, safflower). Also useful are many of the flowering oilseed crops such as sesame and mustard.

• **Soil fumigants or nematicides**: Marigolds, sesame mulch and roots, nasturtium, many *Crotalaria* species, castor bean plants, tamarind roots, custard apple roots, etc. The mulch acts as a host to predatory fungi and also suppresses weeds.

Such crop guilds are rarely attacked by insects. The occasional crop that *is* heavily attacked can be left alone to build up predators; however, a loss of one crop represents very little loss in the total yield of all the crops. All gardeners know of this occasional loss due to seasonal effects, and also of special good seasons producing heavy yields.

Hedgerow, along with weedy headlands, unmown roadside verges, ponds, stone-filled hollows, piles of old timber, mulch-filled holes in the ground, and fallen logs left on borders harbour many species of predators such as frogs, birds, lizards, dragonflies, etc., which assist in keeping pest numbers to a minimum.

Common Interplants of the Deccan

A three-crop standard is of sorghum, cowpea, and pigeon pea, alley cropped in rows 2 metres apart. Sorghum is harvested first, and the dried stalks are stored as fodder. Pigeon pea is harvested October and November, and can be coppiced if perennial; the tops are added to rows with sorghum straw and cowpea vines. Sunflower is often planted on the edges of the field. It is possible to sow oats or wheat as a winter crop between pigeon pea rows, thus giving a four-crop sequence, which could be done if maize replaces sorghum.

A common random scatter of flowering crop contains coriander, celosia, safflower, and fenugreek. Rows of flax can be planted through this crop, along with black sesame seed. Occasionally fennel or dill is added. Such a crop is alive with flowers and insects in mid to late November. Celosia is more of a scattered weed, harvested for buffalo fodder at times, and is the dominant weed of abandoned fields in this period. Mung bean may also be harvested in late October-early November.

Another three-crop mix is sugar cane, with sesbania dominant and a turmeric understorey. Here, sugar cane is the main crop and is irrigated. In October the cane is tied in bundles to allow more light to the turmeric, or every 3 years it is cut and the sesbania left in the field or cut for poles or forages. A variation is where turmeric

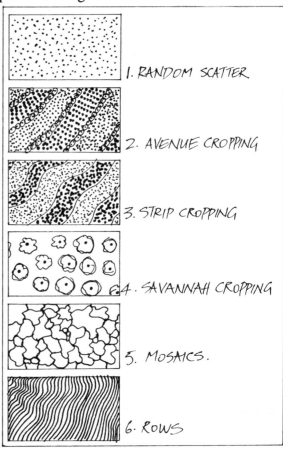

FIGURE 6.12 Different types of geometrical patterns for mixed cropping.

is the main crop, and scattered sesbania and castor oil plants are scattered throughout the crop, so that it looks like a low savannah.

Edging mosaics of taller sunflower, castor bean, or strips of maize or *Sesbania bispinosa* can shelter small or more wind-prone crop. Hedges at 30-50 metres supply many yields and functions to the crop. **Figures 6.11** and **6.13** illustrate fields with hedge, windbreak, avenue crop, and swales.

Geometry of Crops in Monsoon Areas

The ways in which soils are ridged and shaped to prevent water runoff and subsequent erosion is important in subtropical to tropical agricultural systems. Many hill farmers use terraces, swales, and ditch and bank, while flatland (less than 3% slope) farmers can adopt simple scatters of mixed seed. Some of the main geometrical planting methods of polycultural fields are given in **Figure 6.12**.

Soils are often ridged (to 20cm high at the

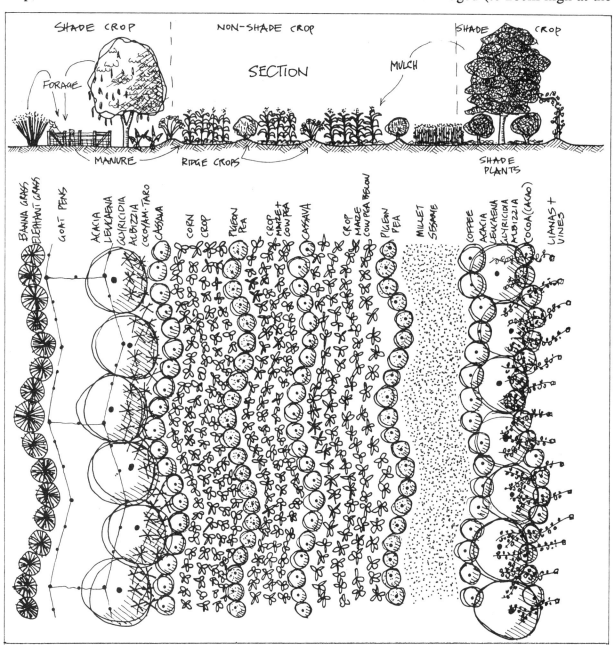

FIGURE 6.12 Nigerian polyculture for the humid tropics; includes pig or goat pen and forage crops. Strip plantings are on contour, with zero water runoff.

141

walls) in oblongs; these are called "waffle-iron" fields, in which each little oblong is only 2x3 to 3x4 metres, so that during rainy periods no runoff can occur. Even one off-season or winter rain can grow vegetables and pearl millet crops using this method of field configuration.

Obviously, all these systems can be combined. Flax and sunflower in strips can lie between pigeon pea avenues 2-3 metres apart, and in these may be planted some large legume trees or field trees at about 40 per hectare. Some strips are planted to random scatters of up to 5 species or more of flowering crop and "encouraged weeds" such as *Chenopodium* and *Amaranthus* spp. Rows of grains lie between avenues of pigeon pea.

6.4
ON-FARM FUELS

Fuel as methane can not only be derived from animal manures, but also from leaf litter and branches under mature forest. Chipped leaves and branches are processed through a biogas digester to produce methane for cooking, heating, and vehicle needs. All waste products, however, should be returned to the forest as nutrient for further growth. For a full explanation of such a bio-energy system, refer to *Another Kind of Garden* by Ida and Jean Pain (see references at the end of this chapter).

For liquid fuels, species yielding sugars for conversion to alcohol (toddy palm, carob, fruit trees) are planted. The tree itself is not cut down, rather the sap (palm trees) or fruits are gathered. Low or no-tillage grains and starchy root crops, of sugar-rich carob beans, plums, sugar cane and beets can all be fermented to alcohol fuel. After fermentation, waste products are returned to the farm in the form of mulch, stock feed, and soil additives. No critical materials are lost, rather all products not directly used for fuel are recycled via animal feed (pig, worm, fish) to plant food, thus cycling nutrients on the farm.

About 5-10% of farm land devoted to fuel production would provide fuel self-sufficiency,

with some surplus. Less area would be needed if sugar-pod-producing tree crops are also developed.

The technology is well-known, but the pretence is that we need more "research" to develop this in Australia. Hogwash! Sixty percent of Brazil's vehicles run on alcohol, and thousands of U.S. farmers now use on-farm stills. These are especially important as energy costs spiral upwards. Perhaps the best argument for alcohol fuel is that the insidious lead pollution from car exhausts is eliminated, thus reducing health hazards in cities. The long-term advantage is that the threat of climatic change due to the burning of fossil fuels and the felling of forests can be reduced or avoided.

Farms and city waste centres are the potential future energy base for essential fuels. With bicycle "freeways" increased and more efficient rail, canal, and sea transport, any society can be self-sufficient in essential transport needs.

The problem is the centralisation of power in large utilities. Vast sums are spent on advising people to "save petrol", whereas the same amount spent on the low-cost distillation plants that would make a community or small town self-sufficient is "not available". The intention is obvious: we are expected to stick with petrol or gas products, lead and pollution, until the oil companies gain control of alcohol fuels. Sometimes one can be pardoned for thinking that we are all crazy, or dumb, or that there is a gigantic conspiracy to keep people down and out. I am inclined to think both factors are operating.

6.5
COMMERCIAL SYSTEMS

For commercial orchards, grain and seed crop, market gardens, and small animal systems (poultry, pigs), small areas of 5 acres or less work better than large acreages devoted to single or even double cropping. It is impossible to completely mulch, water, maintain *and* raise a large variety of plants and animals for multiple functions and multiple yields over a large area (as can be accomplished on a Zone I or Zone II level). Extensive systems, therefore, tend to simplify.

However, this factor can be overcome by a "commonwork" model, where families or groups agree to divide up the work and the products, so that one is responsible for the orchard, while another grows green crop beneath or runs poultry. Someone else might bring bees in during flowering for pollination (and honey production), and manage the firewood crop interplanted with the fruit and nut crops.

Smaller systems are usually easily managed by a farm family with seasonal helpers, and provide high yields due to mixed cropping and intensive management.

Some rules for cash crops are as follows:

• Choose a crop with low bulk (nuts, berries, oil, honey) which cuts down on transport costs.

• Choose a crop suitable for small-scale processing, which reduces the size of the product, prolongs shelf life, and returns a better profit (for example, sale of blackberry jam rather than the blackberries themselves).

• Market your primary produce in (1) organic markets, or (2) special products markets such as delicatessens and restaurants (for truffles, herbs, shiitake mushrooms).

• Grow or produce non-perishable products (grains, nuts, honey, firewood) for sales throughout the year.

• Minimise your costs by using waste products, and by harvesting any unused trees in the district.

• Grow crops in reasonably marketable quantities; also try a few little-known crops or products to experiment with local market acceptance (tamarillos, pepinos, feijoas).

Selling strategies include: direct selling locally at markets or at roadside stalls; self-pick sales; market cooperatives; mail-order catalogues and through subscriber networks (producer-consumer cooperatives where the farmer grows crops by agreement with a town/city consumer group). This strategy began in Japan and is now gaining popularity in the United States where families pay $20 per week in advance for seasonal fruit and vegetables; growers deliver a range of up to 50 products each week to their door.

Some suggested occupations and products are as follows:

Aquatic and edge plant nursery, including fish forages, insectary species, and marshland perennials for bee fodders, duck forage, and wildlife refuges. Also sale of edible and ornamental waterplants, e.g lily, lotus, water chestnut.

Berry fruit and vine nursery, especially in temperate areas, with plants for sale, self-picking service, arbour design plans.

Specialty nurseries, with hard-to-find edible and other useful permaculture plants (tagasaste, honey locust, feijoa, tamarillo, cardoon, autumn olive, comfrey, winged bean, etc.). Also bee fodder plants, and bird, butterfly, and insectivorous insect attractants.

Seed company, gathering, growing, and selling useful and unusual seeds; can be combined with nursery above.

Unusual or useful animals, e.g. silky bantams for gardens, weeder geese, silkworms and earthworms, draught horses, milking goats or cows, specialty goats or sheep (for fine wools), and quail for greenhouses. Can also have rent-an-animal services (chicken or pig tractoring and manuring, sheep and geese lawnmowers, blackberry-eating goats).

Hedgerow and tree species nursery, specific to local region; includes forestry trees for regeneration of farm forests, windbreak trees, forage species for animals, pioneers, bamboos, and selected high-value tree crop species.

General farm crop of organic fruits, nuts, vegetables, eggs, milk, sheepskins, firewood, fresh meats, aquaculture products, cut flowers.

Processed farm crop for higher income (but more effort), such as smoked fish and meats, dried fruit, jams, pickles, feathers (goose down and peacock plumes), dried flowers (bouquets, wreaths).

Craft supplies from coppiced willow, birch, cumbungi, and bamboo. Also natural dyes from barks, flowers, and fruit.

Insecticidal preparations, such as ground white cedar tree leaves and berries; also sale of insecticidal plants (e.g. garlic, tansy, yarrow, pyrethrum daisy, *Tagetes* marigold, crotalaria).

Herbal preparations, such as natural shampoos and soaps, skin care, comfrey and other medicinal ointments. Also herbal teas (chamomile, raspberry leaf, lemongrass, hibiscus, mints).

Accommodation: health farm, holiday farm, summer camp, venue for workshops and courses.

Teaching and consulting in permaculture systems, a career that starts out locally and may take you globally!

There are many, many more livings which can be made just from intensively and efficiently using even a small amount of land. All it takes is initial planning, some capital, and imagination.

6.6 REFERENCES AND FURTHER READING

Breckwoldt, Roland, *Wildlife in the Home Paddock: nature conservation for Australian farmers*, Angus & Robertson, 1983.

Dept. of National Development, *The Use of Trees and Shrubs in the Dry Country of Australia*, Forest & Timber Bureau, 1972. (Use of trees in soil conservation, forestry, livestock fodder, honey production).

Douglas, J.S. and Robert A. de Hart, *Forest Farming*, Watkins, London, 1976.

Fukuoka, Masanobu, *The One-Straw Revolution*, Rodale Press, Emmaus, 1978.

Fukuoka, Masanobu, *The Natural Way of Farming*, Japan Publications, Inc., Tokyo & New York, 1985.

King, F.H., *Farmers of Forty Centuries: permanent agriculture in China, Korea, and Japan*, 1911, Rodale Press, Emmaus.

Logsden, Gene, *Small-scale Grain Growing*, Rodale Press, Emmaus, 1977.

NSW Forestry Commission, *Trees and Shrubs for Eastern Australia*, NSW University Press, 1980.

Pain, Ida and Jean, *Another Kind of Garden*, self-published in France, 1982. Available from Biothermal Energy Center, PO Box 3112, Portland, ME 04101, USA.

Reid, Rowan, and Geoff Wilson, *Agroforestry in Australia and New Zealand*, Goddard & Dobson, Box Hill, Victoria 3128, 1985.

Smith, J. Russell, *Tree Crops: a permanent agriculture*, Devine-Adair, Old Greenwich, 1950.

Snook, Laurence C., *Tagasaste (Tree Lucerne) High Production Fodder Crop*, Night Owl Publishers, Shepparton, VIC 3630, 1986.

Turner, Newman, *Fertility Pastures and Cover Crops*, 1974. Available from Rateaver, Pauma Valley, California 92061
(Valuable guide for temperate herbals leys and biological agriculture).

CHAPTER 7

Animal Forage Systems and Acquaculture

"You don't have a snail problem; you've got a duck deficiency!"   Bill Mollison

7.1 INTRODUCTION

In considering a permaculture as a complete ecosystem, animals are essential to control vegetation and pests and to complete the basic nutrient cycle of a farm. Despite their inefficiency in protein conversion, their diverse products make them invaluable. **Figure 7.1** shows the needs, products, and functions of animals in the system.

In essence, animals can be used as:
• Providers of high quality manures.
• Pollinators and foragers, collecting dispersed materials from a permaculture.
• Heat sources, radiating body heat for use in enclosed systems such as greenhouses and barns.
• Gas producers (carbon dioxide and methane), again for use in enclosed systems such as greenhouses and methane digesters.
• "Tractors", which dig soils. Poultry and pigs are efficient soil-turning, weeding and manuring "machines" for enclosed spaces.
• Draught animals operating pumps and vehicles.
• Pioneers for clearing and manuring difficult areas prior to planting, e.g. goats in black-berry patches.
• Pest control mechanisms, devouring pupae and eggs of pests in fallen fruit, or in trees and shrubs.
• Concentrators of specific nutrients, such as nitrogen and phosphates from flies and wasps.
• Cleansing filters for water (e.g. mussels).
• Short grazers aiding in fire control.

Vegetarian communities are still able to use animals (one-sex or sterilised populations) as providers of fibres, eggs, and milk; as grazers for fire control; and as manure providers for gardens and orchards.

In permaculture systems, a range of animal feeds (fruit, foliage, pods, nuts, seeds, and tubers) are planted so that animals self-forage, taking most of what they need from the natural world, and at the same time manuring, controlling vegetation and pests, and converting plants to protein. Animals in a free-range system will put on weight more slowly than when fed concentrated feeds, but fat accumulation is less and fats are soft and unsaturated. The diversity and regularity of free range diet is basic to animal health.

To lay out important forages, we must study the needs and characteristics of each animal and plan our plant system accordingly (i.e. chickens are scratchers, geese are grazers, and pigs are rooters). The following sections give a brief overview of several important animals, including their needs, characteristics, and products.

The following small animals can be placed in any appropriate zone, according to their popula-tion. Rabbits, pigeons, and quail are generally close in (Zone I or II), while other birds can range from Zone II to IV.

■ RABBITS

Rabbits supply both manures for the garden and meat for the table. They are grazers and browsers, and will eat grass, soft vegetation and twigs, and selected household scraps. They bur-

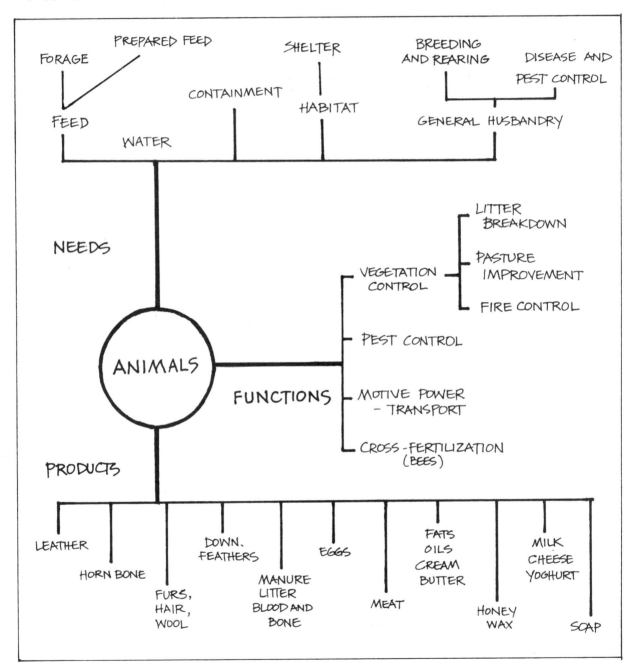

FIGURE 7.1 Schematic of animals in permaculture.

row into the ground and will cause damage to soils and vegetation if not properly enclosed. They yield fur (angora rabbits produce a valuable hair which is combed out periodically for sale or home use), meat and manures.

When penned above worm boxes, their droppings are turned into rich compost (**Figure 7.2**). Or, rabbit pens lead to runs planted with such forage crops as lucerne (alfalfa), tagasaste, and clover. Rabbits can also be placed in the garden to eat grass in a moveable cage between rows.

■ PIGEONS AND QUAIL

Pigeons are kept all over the world and valued for their phosphate-rich manures. They are caged above ground and their manures swept out from underneath, or dovecotes are built and manures and squabs periodically collected (**Figure 7.3**). Pigeons eat seeds and grains, which can be grown and harvested from the garden (corn, sunflower seeds, peas, wheat). They provide eggs and squabs.

Quail, in Japan, are an integral part of small-scale farms, providing eggs and meat and need-

FIGURE 7.2 Rabbit manures fall through mesh floor to earthworm boxes below. Hessian cloth and sprinklers keep rabbits shaded and cool in hot climates.

ing little attention. As they are insect-eaters, they do not harm garden vegetables and may be placed in the greenhouse to great advantage (as

FIGURE 7.3 Dovecotes can be of diverse materials: (**A**) mud (**B**) wood (**C**) brick.

147

long as they are able to move outdoors during hot summer months).

■ GUINEA PIGS

Guinea pigs, an important source of protein in some South American countries, are kept very close by the house (or may actually live *in* the house), and fed garden scraps and seeds. They are useful in weeding around small trees, either through a large-mesh cage, or even on their own (they must be given a small house, or shelter, however, to protect them from hawks).

■ DUCKS

Ducks are excellent permaculture animals and have many advantages. They can be raised without elaborate housing, and will readily thrive on natural foods. They clean up waterways of green algae, water weeds, and tubers, at the same time fertilising watercourses which aids in fish and eel production. They eat insects, and slugs and snails in orchards and gardens, and because they do not scratch or eat mature greens, can be let into the garden at appropriate times to consume insects. **Caution**: they will destroy small plants with their feet; also some duck species (Muscovies) are vegetation eaters, although they confine themselves mainly to grasses.

Because ducks do not scratch mulches, they can be ranged in mulched gardens and orchards. Ducks will also lay 98% of their eggs before 10 a.m., so they can be let out early to range; they respond to routine and come home at night to the pen (but must be trained for this with handfuls of grain).

There are a few disadvantages, in that they do not readily eat many of the scraps chickens can consume, and will turn a small pen into a mudbath unless the ground is sandy or easily-drained, or covered with 10-15cm of coarse pea gravel and located upslope.

Duck foods include:

• Meat: water crustaceans, slugs, snails, grubs, larvae.

• Greens: wilted comfrey, clover, lucerne, dandelion, succulent grasses.

• Waterplants: *Azolla*, duckweed (*Lemna*); water ribbons (*Triglochin*); sweetgrass (*Glyceria*) and wild rice (*Zizania aquatica*).

• Tree-food: pin oak, cork oak, swamp holly, water elm, mulberry.

• Grains: corn, oats, wheat (preferably hammer milled or ground, or soaked for several days until soft and partially sprouted).

Figure 7.4 illustrates ways in which ducks can lay eggs unmolested by foxes, goannas, or snakes.

■ GEESE

Geese are cheap to feed; living on grasses (bermuda, nut grass), clover, lucerne, and various weeds such as ragweed. They find many broad-leafed plants unpalatable, and so have

FIGURE 7.4 Protection from predators (foxes): (**1**) covered cage, opening in water (**2**) island with pampas grass, bamboo, hollow logs (**3**) shallow, marshy area with cumbungi fringe.

been used to control grasses in commercial crops, waterways, and lawns. They will weed strawberries, tobacco, cotton, mint, asparagus, corn, sugar cane, sugar beet, flowers, grapes, fruit orchards, nut groves, and tree nursery areas. They manure fields and orchards without scratching away mulch. They work seven days a week, without pay, vacations, or strikes. Who could ask for more?

Geese can also be used as "watchdogs" as they set up a noisy alarm at a stranger's approach. They have even been trained to herd sheep. Other advantages are their eggs, meat, and goose feathers.

Geese need careful management if being used as weeders in crops or orchards as their feet will destroy small plants and they will eat ripening fruit. Also, although they are excellent lawnmowers, they prefer pasture that is short and succulent, so areas may need to be mowed in spring once or twice during prolific growth.

■ BEES

Bees are very useful in the garden and orchard as pollinators. Their products are honey, pollen, and beeswax, and their needs are water and a constant source of nectar (flowers). To keep bees on site all year round, a complete forage system must be planned for each month. However, flowering and yields of nectar varies greatly from year to year, depending upon weather conditions, so at times bees are fed sugar water or the hives are moved some miles away to another nectar source.

Bee forages to consider are native vegetation and pasture species such as clover and lucerne; orchard trees (apple, cherry, almond, peach, plum); berry bushes; and herbs (lavender, bergamot, borage, comfrey). A mixture of these will ensure an almost-constant supply of nectar, except in areas experiencing severe winters (snow).

7.3
POULTRY FORAGE SYSTEMS

Whenever possible, Zone II should include the range of some high manurial animals like chickens, and they should be housed at the edge of Zone I, or very close to it. Here we can exploit a larger system (Zone II) to enrich a smaller one (Zone I), through the use of an animal converter.

Chickens, besides their direct products of eggs, meat, feathers, and manure, also eat insects, greens, and fallen fruit. They scratch an area clean if confined in a small space, and can be used to patrol a fenced boundary area (e.g. between garden and orchard) to keep weed species from invading the garden). This scratching feature is especially useful for fire control in the fire sector.

Although poultry do need care and maintenance, the permaculture system is designed so that chickens range to feed and take care of themselves. Therefore, we need to carefully plan a chicken forage system that accommodates the needs of the chickens and uses its products.

Strawyard

The strawyard is a small area connected to the hen house containing productive trees, bushes, forage plants, and spiny shelter for raising young chicks. This is either planted before chickens are introduced, or protected from them in early years. To protect trees, a rough mulch of twigs or stones can be used, with medium-mesh screen holding the mulch from being scratched away by the chickens. The strawyard itself is continually heavily mulched with straw, sawdust, corn stalks, hedge clippings, wood shavings, small branches, pine needles, leaves, weeds, and bark. With the strawyard bordering the garden, greens and shrub clippings can be thrown over the fence for the chickens.

The strawyard opens to various pens or "runs" which have been planted in succession with greens, grains, roots, and fruit. Chickens are rotated either on a seasonal basis, or whenever the vegetation is ready (**Figure 7.5**). Additionally, the strawyard can open on to Zone II and Zone III forage systems.

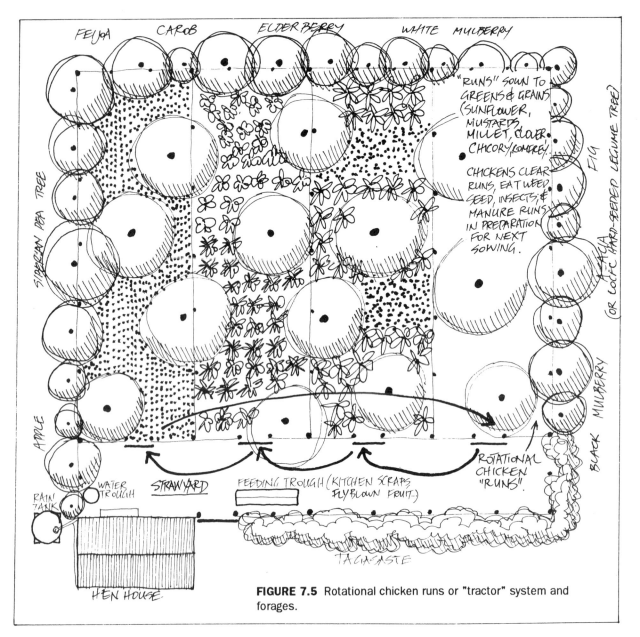

FIGURE 7.5 Rotational chicken runs or "tractor" system and forages.

Plant Species

Useful plant species should be developed to suit the climate, as well as available water and plants. Such a list should include plants which provide:

• Armoured and spiny shrubs for protection of chicks from predators (usually hawks), e.g. *Prosopis juliflora*, *Acacia armata*, boxthorn or any locally-adaptive thorny plant.

• Fruits that can be eaten as they ripen and fall from trees and shrubs, e.g. mulberry, boxthorn, taupata, elderberry, passionfruit.

• Grain foods, e.g. corn, millet, wheat, buckwheat, oats, beans and peas, pigeon pea, taupata.

Many grain and seed foods can be collected and stored for the winter months, when feed is thin on the ground. These include acorns, sunflower seeds, corn, and carob pods.

• Seed foods, such as tagasaste, sunflower, amaranth, acacias, black locust and honey locust, pea tree, salt bushes.

• Greens. Chickens will eat *any* young greens, including garden greens, some comfrey, lespedeza, lucerne, buckwheat, cleavers, pea shrub, young grass, parsley, etc.

• Other. *Household scraps*, excluding citrus peels, coffee and tea grounds, onion skins. *Minerals*: grit, ground eggshells, bonemeal, cinders, crushed oystershell. *Medicinal herbs*: garlic, wormwood, chopped nettles.

In addition, chickens need protein in the form of insect foods. A termite and slater (pillbug) trap can be constructed by arranging old logs in the strawyard and turning them over occasionally for a chicken feast. Rolled-up newspapers placed in trees and shrubs in the evening are shaken out the next morning into the strawyard for a chicken feast.

BLACK WALNUT
WALNUT
CHESTNUT
OAK
PEARS
MULBERRIES
HONEY LOCUST
CRAB APPLE
CAROB
OLIVE
TREE LUCERNE
PEATREE
COPROSMA
BAMBOO

COMFREY
SUNFLOWER
LUCERNE
BUCKWHEAT

FIELD ROOST

ZONE III

ZONE II

MULCH

STRAWYARD

ROUGH COMPOST

GLASSHOUSE

(HOUSE)

ZONE I.

CHICKEN-TRACTOR BEDS.

FIGURE 7.6 Rural chicken forage layout, showing all possible components (greenhouse, strawyard, and access to gardens and orchards). Note moveable chicken "cage" over garden beds, so that chickens do not range freely over the garden.

WARM/TEMPERATE SPECIES LIST FOR POULTRY FORAGE

The following list is certainly not exhaustive.
There are many more local or native species in your own area which can be added.

Species with seeds and pods in summer

Tagasaste: Early to mid-summer seed drop. Foliage edible, also for sheep, cattle, goats. Leguminous nitrogen fixer.

Siberian pea shrub: Poultry fodder and predator cover; seeds also edible. Used as windbreak, ground cover, bee forage, and soil builder (a legume).

Honey locust: Seeds and pods stored for milling. Also windbreak, fodder for larger animals. Also black locust for seed (leaves may poison larger stock).

Acacias such as *A. albida, A. aneura, A. victoriae*, etc. for hard-seeded species. Acacias make good windbreaks, are nitrogen-fixing, and leaves are fed to stock.

Tree and shrubs yielding nuts for storage (autumn-spring)

Black walnut and Persian walnut: Nuts can be stored for 12 months. Also a valuable timber tree, windbreak.

Chestnut: Stored for 6 months only, unless chilled or dried in the sun.

Oaks: Almost all acorns are edible for poultry. They are easy to collect and store either in damp earth, dried, or fresh for short periods of the year.

Berries and fruits yielding flesh or seed (late summer-mid winter)

White and black mulberries: Important poultry food of high protein value. Also elderberries.

Boxthorn: Thorny hedges with berries and seeds eagerly sought by poultry. Wind tolerant.

Taupata: A useful and hardy set of New Zealand plants for coasts, swamps, understorey, shelter plants. Most are dioecious and need around 5% male plants. Almost all grow from cuttings. Stock like the foliage, which is also of good manurial value. Trees prune well to hedges handy for any chicken-rearing topiarist!).

Amelanchier spp.: Provide a range of berry fodders (e.g. serviceberry); also hawthorns (*Craetagus*) and *Elaeagnus* spp. (autumn olive, Russian olive). These plants provide thorny hedge for protection of chicks.

Tamarillo: Short-lived tree-shrub matures in 2 years, yielding copious amounts of tasty fruits. Other *Solanum* species include kangaroo apple, pepino, tomato, cape gooseberry, and sodom apple, all very good chicken forages.

Vines for Fences and Trellis

Passionfruit: Most passionfruits are tropical and subtropical; however, banana passionfruit (*Passiflora mollisima*) can tolerate light frost.

Choko (chayote): A climbing perennial vine yielding large green vegetables. Is rampant in the tropics, and can be used to cover areas of noxious vegetation, e.g. lantana.

Dolichos spp.: Annual and perennial beans; species range from temperate to tropical, evergreen perennial to annual.

Greens and Seeds as Herb Layer

On extensive free range, an herbal layer of clovers, medics, lucerne, chicory, and fennel may be sown, along with mixed grasses. Ducks and geese also appreciate the seed-heads of rye grasses and clovers. Pokeweed is eaten by birds, especially pigeons. Also grown can be millet, lupin species, perennial buckwheat.

Species for Broadcast Sowing in Rotated Strawyards

Sunflower: Greens eaten; heads stored in autumn for winter food.

Millets, corn, buckwheat, and the usual grains of wheat, rye, barley, oats, etc.: sown in rotation so poultry get small greens, some saved for winter seed. Also legumes such as field peas.

Amaranth: Wide range of tolerances; seed grains suited to poultry. Also quinoa.

Herbs, Weeds, and Throw-over Crop

Shepherds purse: This herb is an excellent poultry forage and has a beneficial effect on egg production. As it is usually a nuisance in areas where it is not wanted, poultry are a valuable control. Also chickweed (poultry eat the seeds).

Cleavers: Another "weed", it is a useful seed and greens plant for poultry, containing important sources of iron and iodine. For free-range poultry, plantings of cleavers may need to be protected by brush or netted fence enclosures.

Chard or silver beet: An easily-grown garden plant which can be over-sown especially for poultry and thrown from the garden over the fence to the strawyard.

Chickens ranged in the garden under controlled conditions "tractor" an area and leave it completely manured. Permanent or portable structures (enclosed by chicken wire) are designed to fit garden beds or garden areas with chickens allowed in after a harvest and before re-planting. This usually works only for broad beds where a crop is harvested all at once, rather than the small pathside beds close to the house. Bantam chickens are small and eat mainly insects, cutworms, and slugs, leaving grown vegetation alone.

Figures 7.6 and **7.7** give some ideas for chicken forage forests, centred both on a homestead and on the suburban backyard.

Figure 7.8 shows a chicken-heated greenhouse, which is a self-regulated structure. In winter, the greenhouse, via vents, heats the chicken-house (and the heat from the bodies of chickens maintains a warm temperature in the greenhouse), while in summer, vents are closed, and chickens spend most of their time foraging outside. The two sides are screened off from one another, but contain a door or other access to collect eggs from nest boxes and to feed chickens any greens from the hothouse. Chickens provide carbon dioxide and feather dust to the greenhouse, along with manure/litter which is eventually composted.

■ TROPICAL CHICKEN TRACTOR SYSTEM
Following is a sample system developed by Dano Gorsich of Molokai, Hawaii. The system itself is not confined to the tropics and with modification can be adapted to temperate regions and even drylands if there is a plentiful source of water. Plants do not grow as quickly in these climates as in the tropics, so adjustments

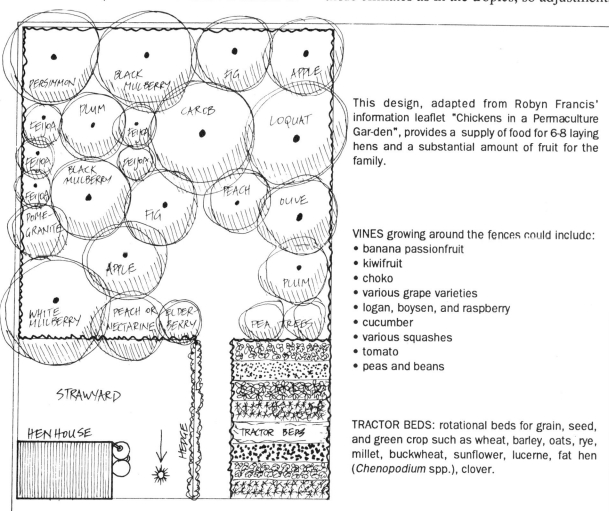

This design, adapted from Robyn Francis' information leaflet "Chickens in a Permaculture Garden", provides a supply of food for 6-8 laying hens and a substantial amount of fruit for the family.

VINES growing around the fences could include:
- banana passionfruit
- kiwifruit
- choko
- various grape varieties
- logan, boysen, and raspberry
- cucumber
- various squashes
- tomato
- peas and beans

TRACTOR BEDS: rotational beds for grain, seed, and green crop such as wheat, barley, oats, rye, millet, buckwheat, sunflower, lucerne, fat hen (*Chenopodium* spp.), clover.

FIGURE 7.7 Backyard permaculture chicken forage forest for a Mediterranean-type climate.

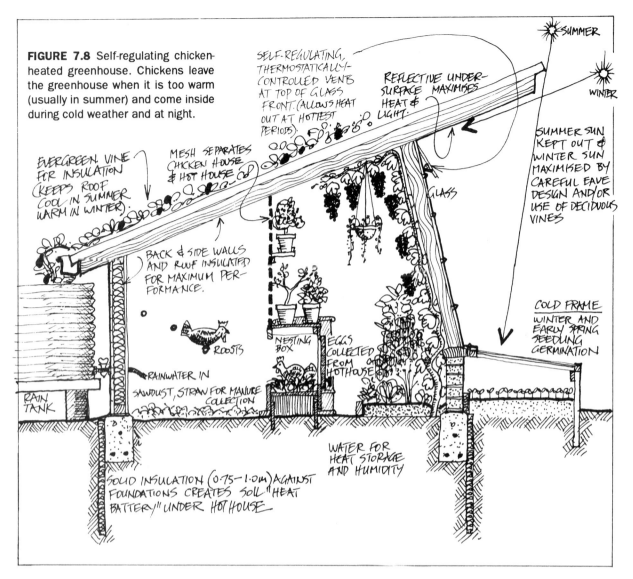

FIGURE 7.8 Self-regulating chicken-heated greenhouse. Chickens leave the greenhouse when it is too warm (usually in summer) and come inside during cold weather and at night.

SELF-REGULATING, THERMOSTATICALLY-CONTROLLED VENTS AT TOP OF GLASS FRONT. (ALLOWS HEAT OUT AT HOTTEST PERIODS).

REFLECTIVE UNDER-SURFACE MAXIMISES HEAT & LIGHT.

SUMMER

WINTER

SUMMER SUN KEPT OUT & WINTER SUN MAXIMISED BY CAREFUL EAVE DESIGN AND/OR USE OF DECIDUOUS VINES

EVERGREEN VINE FOR INSULATION (KEEPS ROOF COOL IN SUMMER WARM IN WINTER)

MESH SEPARATES CHICKEN HOUSE & HOT HOUSE

GLASS

BACK & SIDE WALLS AND ROOF INSULATED FOR MAXIMUM PERFORMANCE.

ROOSTS

NESTING BOX

EGGS COLLECTED FROM HOTHOUSE

COLD FRAME WINTER AND EARLY SPRING SEEDLING GERMINATION

RAINWATER IN

SAWDUST, STRAW FOR MANURE COLLECTION

RAIN TANK

WATER FOR HEAT STORAGE AND HUMIDITY

SOLID INSULATION (0.75-1.0m) AGAINST FOUNDATIONS CREATES SOIL "HEAT BATTERY" UNDER HOTHOUSE.

must be made.

To prepare an area of 0.2 hectare (1/2 acre), divide the area into 5 pens of about 10 metres by 6 metres or so. Stock with about 50 chickens (layers) in one pen until all grasses and weeds are grazed off and removed. (Pens can be laid out as in **Figure 7.5** so that only one henhouse with laying boxes need be used.) Add a scatter of lime, move the chickens to the next pen, chip or rake over the ground in the first pen, and plant vegetable crop (melons, Chinese cabbage, tomatoes, etc.). Also plant leucaena or other legumes just outside the pen, along with 15 or so papaya seedlings or bananas.

Each pen has a small roost and nest box in it, which can be moved around to all pens, and

water and food are supplied.

After the chickens clean the second pen (6-10 weeks), the first pen is harvested, and replanted with root crops. The second pen is planted like the first. In some of the pens, important tropical fruit and nut trees are planted.

After the chickens clean the third pen, the second pen is harvested (10 weeks). The first pen is dug for roots, the third pen is planted to green crop (peas, beans, brassica)...and so on, for the rest of the pens.

The chickens are returned to the first pen after the root crops are out and fruits and trees well-grown or adequately protected. This pen has been sown to buckwheat, sunflower, pigeon pea, rice or barley 10-12 weeks before the chick-

1	2	3	4	5
CHICKENS IN GRASS AND WEEDS	PEN LIMED AND SOWN TO GREEN VEGETABLES	VEGETABLE HARVEST, ROOT CROP PLANTED, FRUIT TREES PLANTED	ROOTS HARVESTED, GRAIN CROP SOWN.	GRAIN HARVEST, TREES PROTECTED CHICKENS REINTRODUCED

FIGURE 7.9 Schematic of a chicken tractor system for the tropics. Can be modified for other climates.

ens are returned. Grains and seed-heads are stored in bundles hung under a roof, and these are fed out to the chickens as needed, along with papaya and banana. Leucaena seeds fall into the pen. **Figure 7.9** shows the rotational sequence of one pen.

After one year, the chickens can self-feed on grains, crop wastes, and papaya. They can also be let out of a pen every day to forage greens. If fruit trees are thickly planted and shade out the vegetable pens, the system can be expanded on fresh grassland (next to the previous pens) which is then retained for vegetable, root, grain, and scattered fruit crop. After two years, an area of 1 acre is in good production. We see here a combined system, using chickens as work units and producers. Pigs could as easily be used.

7.4
PIG FORAGE SYSTEMS

Pigs are forest and marshland foragers, and they like to graze, forage, and root (dig up tubers and roots). They will graze all grasses, herbage, and rambling vines; forage for fallen fruit and nuts (mulberries, persimmons, figs, mango, carob, acorns, avocados, etc.); and dig up yams, potatoes, bamboo, arrowroot, bracken, and sunroot.

Pigs on range are healthier, cheaper to feed, and have less saturated fats than pigs kept in sheds. They are not always suitable for bacon, however, and may need grain-feeding for 2-4 weeks to harden (saturate) the fats. Winter shedding may be necessary in cold winter climates,

and a farrowing pen is needed for a sow and her litter.

Pigs are most cheaply kept where some dairy, orchard, root crop, or meat wastes are available, and do well on restaurant or household food scraps. Good range pasture is of legumes (clover, lucerne), comfrey, chicory, and young grasses. Pigs will eat 11 kg wet weight of this material per day, and have larger appetites than confined pigs. They also need seed, fruit, or kernels.

To prepare a free-range planting area, the ground should be ripped (not ploughed) and limed, then sown down to good grass legume mix, with comfrey, sunroot, and arrowroot pieces pushed into the rip lines. Trees can be planted just outside fences and in corners protected by electric fencing. Any fruit trees are useful, and pigs are beneficial in *mature* orchards.

In a large system, 20 pigs per 4000 square metres (1 acre) will "plough" (by scratching and rooting) the area for planting comfrey, sunroot, lucerne, chicory, and clover. It then needs to rest. Pigs will remove gorse, blackberries, and small shrubs. They can be followed by sowing to pasture, then cattle, then pigs again.

It take 3-5 years to develop a full complement of foods on range, and even then some of this must be thrown over the fence to the pigs, as is the case for bananas and papayas, for pigs can destroy young trees.

Figures 7.10 through **7.12** show sample pig systems.

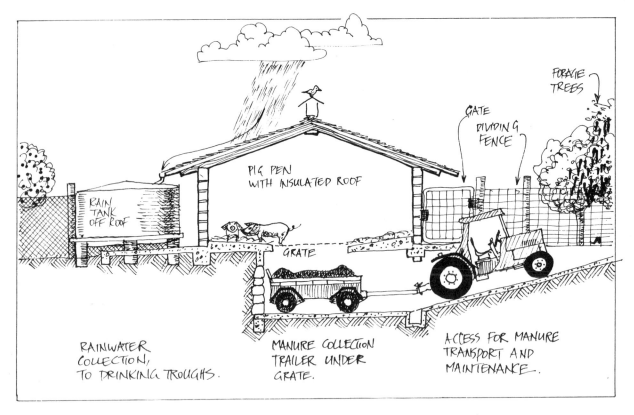

FIGURE 7.10 Cross-section of pig pen with manure collection arrangement.

FIGURE 7.11 View of pig grazing system with tree forages and fence arrangement to keep pigs from damaging tree roots.

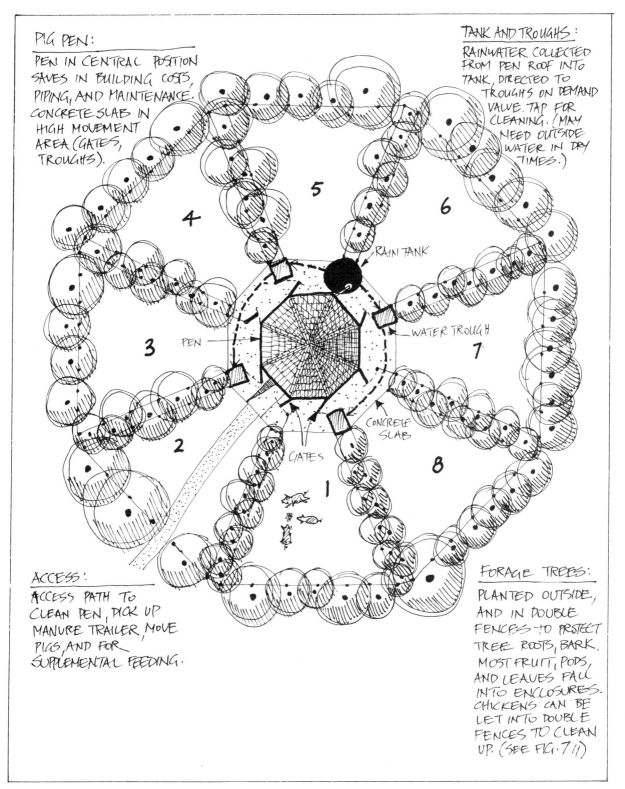

PIG PEN:
PEN IN CENTRAL POSITION SAVES IN BUILDING COSTS, PIPING, AND MAINTENANCE. CONCRETE SLAB IN HIGH MOVEMENT AREA (GATES, TROUGHS).

TANK AND TROUGHS:
RAINWATER COLLECTED FROM PEN ROOF INTO TANK, DIRECTED TO TROUGHS ON DEMAND VALVE. TAP FOR CLEANING. (MAY NEED OUTSIDE WATER IN DRY TIMES.)

RAIN TANK

WATER TROUGH

PEN

CONCRETE SLAB

GATES

ACCESS:
ACCESS PATH TO CLEAN PEN, PICK UP MANURE TRAILER, MOVE PIGS, AND FOR SUPPLEMENTAL FEEDING.

FORAGE TREES:
PLANTED OUTSIDE, AND IN DOUBLE FENCES TO PROTECT TREE ROOTS, BARK. MOST FRUIT, PODS, AND LEAVES FALL INTO ENCLOSURES. CHICKENS CAN BE LET INTO DOUBLE FENCES TO CLEAN UP. (SEE FIG. 7.11)

FIGURE 7.12 Rotational grazing system for pigs. Pens may be planted to comfrey, sunroot, lucerne, choko, potato, and other root and green crops. Trees are oak, mulberry, fig, olive, honey locust, carob (dry areas), chestnuts, papaya, and banana (tropical areas). Forage must be closely monitored in order to time moves from one pen to another.

157

7.5 GOATS

Besides their value in milk and meat production, goats are useful for clearing new country. On abandoned pasture with gorse or blackberries, goats can be used to bring areas under control for future planting, either temporarily penned in numbers, or on individual tethers and moved every few days. If milking goats are used for this process, concentrated feed is also necessary for good milk production.

For a small number of goats (1-3) we can develop a pen with mesh fence to 2 metres surrounded by trees and shrubs. For more edge, enclose two rows of tagasaste into the pen itself, as illustrated in **Figure 7.13**. Some trees that withstand limited grazing by goats are weeping willows, mulberries, tree medic, some acacias, leucaena, tagasaste, and elderberry. Goats enjoy acorns and the pods of carob, honey locust, pea tree, and *Prosopis* spp.

Goats are very destructive to cultivated plants, as apart from browsing, they debark trees. Tethering and the use of orchard halters will allow goats into the more delicate parts of the systems for short periods, but goat husbandry in large numbers is incompatible with permaculture.

7.6 PASTURE CROPS AND LARGE ANIMAL FORAGE SYSTEMS

Pasture crops and forage systems for cows and sheep are usually fairly extensive (8 hectares or more will carry enough stock for a modest living, depending on suitable landscape and climate). Although much of the area is sown to grasses and legumes such clover, there is an emphasis on trees within the system to serve the functions of:

• Feed during drought or times when grasses are sparse;

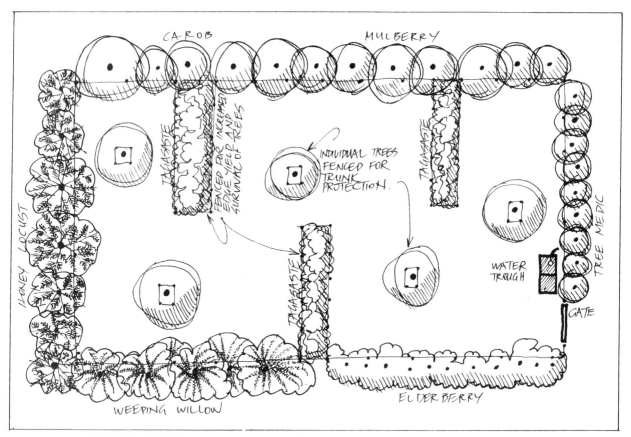

FIGURE 7.13 Milk goat grazing pen and forages adapted from a design by Lea Harrison.

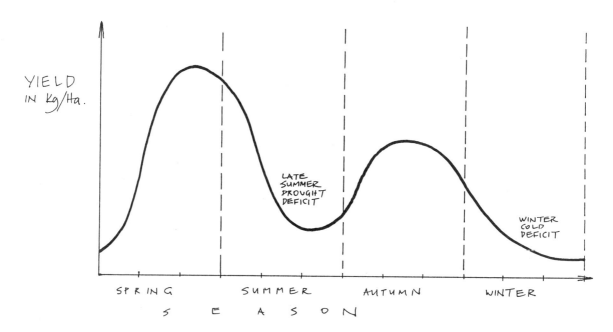

FIGURE 7.14 Growth curve of grasses (for temperate climates) show two deficit periods: winter and late summer.

• Protecting stock from severe wind, snow, rain, and sun (windbreaks and shade-trees);

• Restoring soil fertility to nutrient-depleted soils through leaf litter and nitrogen-fixing legumes.

• Water catchment protection above dams and on steeper slopes (cattle must be kept off these areas); and

• Erosion prevention on slopes and in gullies.

■ PLANNING AN EVEN FODDER DISTRIBUTION

Pasture animals need a source of water, shelter from severe weather, a salt lick, and food, which can be separated into (a) annual and perennial grasses and legumes, (b) sugar pods such as carob and honey locust (summer), (c) carbohydrates as sprouted grain and silage (winter), and (d) plant foliage for an even yield forage over the year.

The age-old problem of a seasonal fodder or forage shortage is illustrated in **Figure 7.14**. In temperate climates, where winter rainfall dominates, both annuals and perennials in pasture reach peak productivity in spring, with a lesser autumn flush of growth if there are early rains. Although the sale of young stock or the culling of herds after breeding reduces the summer feed requirements, it is obvious that there is a shortfall in midsummer and midwinter feed, the former because of summer drought and the latter due to the cold and slow growth of plants.

Tree-crop infills should be planned to take up the gaps that pasture alone leaves. For example,

FIGURE 7.15 Netted tagasaste or taupata. Plant grows through the net and is browsed off. The net can also cover the top of the hedge.

159

some midsummer feed is provided by carob and honey locust pods, the foliage of taupata, pampas grass, and tagasaste, and autumn/winter feed by the same foliage plants plus the great variety of oaks (acorns) and chestnut and black walnut (nuts). Both these types of feed are concentrated, high-energy foods, enabling the more efficient use of dry pasture.

Traditionally, the foliage of kurrajong, willow and poplar has been slash-felled to tide herds over drought. For self-feeding systems under forage forest, plant strips of low forage foliage where herds can be turned in for short periods. Netted hedges of tagasaste have been used to great advantage in New Zealand; cows and sheep cannot destroy the plant, but nibble the succulent leaf growth through the netting if they are turned into the area every month or so during plant growth (**Figure 7.15**).

A gradual (4-10) year change-over to the correct balance of tree crop species would eliminate the need for expensive forage harvesters, feed-grain storage and processing, and haymaking that is an essential part of "pasture only" farming we see today. It also suits the comfort and well-being of animals, which can range into forest when temperatures are extreme, and occupy pastures in the tolerable periods of spring and autumn.

As a secondary effect then, less stress is placed on the herds from heat and cold shock, and far less energy is needed by the farmer and the flock over the whole year. An estimated 15% of beef yield is lost due to lack of shelter alone. Richard St. Barbe-Baker asserted that where 22% of the land is planted to productive trees,

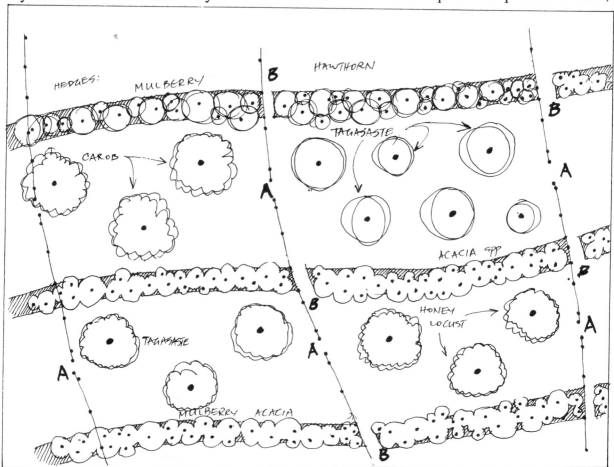

FIGURE 7.16 Rotational paddocks with hedges between fenced strips along the contour. Cows can be rotated along the contour through gates A or between the contours through gates B. Paddocks can be sown to forage crops. Once established, the hedge strips can accommodate periodic browsing from cows. Gates and fences should always be located on the ridges (not in valleys) for this prevents erosion occuring in these heavily-used areas. Adapted from a design by Tony Gilfedder.

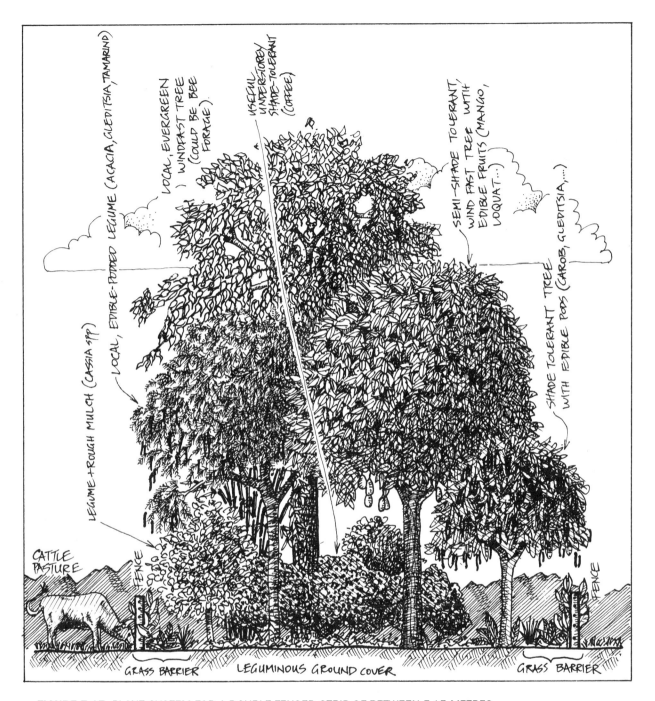

LEGUME + ROUGH MULCH (CASSIA SPP)

LOCAL, EDIBLE-PODDED LEGUME (ACACIA, GLEDITSIA, TAMARIND)

LOCAL, EVERGREEN WINDFAST TREE (COULD BE BEE FORAGE)

USEFUL UNDERSTOREY SHADE-TOLERANT (COFFEE)

SEMI-SHADE TOLERANT WIND FAST TREE WITH EDIBLE FRUITS (MANGO, LOQUAT...)

SHADE TOLERANT TREE WITH EDIBLE PODS (CAROB, GLEDITSIA,....)

CATTLE PASTURE

FENCE

FENCE

GRASS BARRIER

LEGUMINOUS GROUND COVER

GRASS BARRIER

FIGURE 7.17 PLANT SYSTEM FOR A DOUBLE-FENCED STRIP OF BETWEEN 5-15 METRES
Structure: Central row of large trees.
             Tough plants as hedges against fences.
             Delicate plants in central sheltered space.

NOTES: Double-fencing allows a tree fodder hedge in an extensive grazing area.
             The enclosed space can become a sheltered habitat for birds and animals.
             The mixed strip acts as a windbreak.
             The environment inside the strip becomes sheltered and shady, suitable for berry fruits and other
              useful plants if drip-irrigated.
             The strip can provide harvested and foraged yields to large animals next to it.
             Fences can allow passage of smaller domestic species (poultry) on range.
             The strip can be the beginning of a "rolling permaculture" system.

161

yields double on the remaining 78% of the land surface, so that in reality no yields are lost by farm forestry.

To build up surrounding hedgerows, undersow or set plant lucerne, comfrey, chicory, dandelion, with a mid-level planting of tagasaste, Siberian pea tree, taupata, and pampas grass, and a tall overstorey planting of willow, poplar (selected high-value forage cultivars), white oak, chestnut, honey locust, and known desirable woody browse (hawthorn and *Rosa* spp.). Such hedgerows could be designed to occupy 10% of the area per year, until year 4, when 40% of the total area would be broad, complex, contoured hedgerow of deep-rooted shrubs, with tall-tree browse and even high-value timber trees (**Figure 7.16**). After years 4-5, stock such as sheep and some young cows can be let in, timed and observed, to harvest the area. From years 6-8, longer browse times can be permitted, and in emergencies such species as willow and poplar can be cut and fed to animals as drought rations.

A double-fence system is of use in establishing a permaculture hedge or windbreak on an existing pastoral property with cattle or other large animals on range in open country (**Figure 7.17**). Fencelines are obvious sites for windbreak trees, and in the inner zones, stacking stone along fences and planting hedgerows can eventually replace some fences. A dense, mixed hedgerow of spiny shrubs with a low stone wall is virtually impenetrable to most animals.

Hedgerows greatly add to the productivity of the system, and provide fruit, nuts, wood products (e.g. bamboos), animal forage, bee forage, bird habitat and food. They also act as windbreaks and suntraps.

Feed concentrates do have a place in the system for feeding during periods of poor forage, for fattening, and for maintenance of milk and egg production. The tendency to supply only feed concentrates for rapid weight increase should be avoided. Naturally-concentrated foods should come from within the system (honey locust and carob pods, acorns, chestnuts, grains). Although some animals can be fed these concentrated foods unprocessed, a cracking or soaking and sprouting process may be necessary, especially as sprouting increases the quality of some vitamins many-fold. Grains which sprout at moderate temperatures are most suitable: wheat, buckwheat, lucerne, oats, barley, rice, soy beans, mung beans, lentils, peas, chick peas, pumpkin, cress, sunflower seed, fenugreek, sesame seed and rye. (All these, of course, can also be sprouted for human food.) Hay and silage from more valuable forages on site, such as lucerne, can be used as a stored feed for winter months.

The goals of such pasture/forage tree systems is to constantly cycle nutrients from plants to animals and back to the soil via manures and nitrogen-fixing legumes, and to diversify farm products. Tree products such as carob and chestnut can also be more directly converted to sugars, fuels, food additives, flours, and so on. This is of great value when markets for wool, hides, and meat are in flux, and gives the forest farmer a very great advantage over the "pasture only" grower, who is tied to a single market or product.

In a world whose economics are governed by the cost of energy, farmers need to be fully aware of the potential of polyculture. A one-bet system can fail on one factor. As a local permaculture is zoned, so are farmers zoned from market, hence, supply centres. Increasing distance means increasing cost and greater reliance on home production of vital materials, especially manures and fuels. Attention should be given, therefore, to the tree species and animals selected, with respect to local needs and distance to market.

■ ROLLING PERMACULTURE FOR LARGE PROPERTIES

Rolling permaculture is a method of slowly changing from pasture only to a more productive and diverse system. Almost all large properties, of about 20 ha or more, have areas which can be fenced out with little loss to productivity. This is particularly true of steep, stony, eroded, or problem soils, awkward corners, and cold or

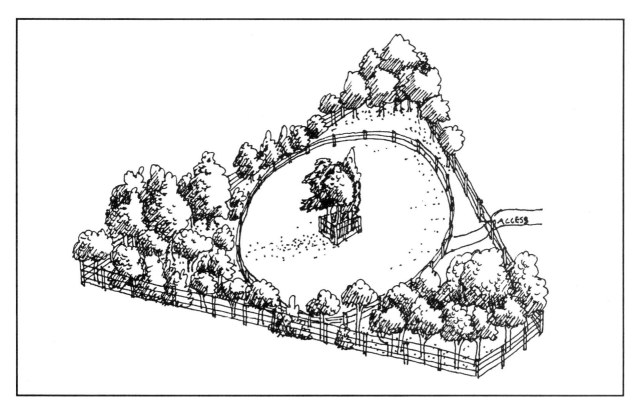

FIGURE 7.18 Rolling permaculture: pasture is electrically-fenced at centres and corners planted to tree crop. Trees species next to fences should be those that can be heavily browsed or are unpalatable to stock.

windswept valleys. We can plant trees which at first provide shelter as hedgerow, and later become a diverse forage and tree crop resource (**Figure 7.18**). The first narrow, or nuclear plantings, contain as many useful species as possible in almost random assembly, fairly thickly planted so that thinnings are available for pole timbers.

The steps to follow for a rolling permaculture are:

1. Exclude animals by fencing, usually electric on a solar charge. Prepare the area by soil rehabilitation (chisel ploughing) and liming, if necessary.

2. Plant a nucleus of trees suitable for windbreak and browse. Mulch and fertilise trees with seaweed solution, blood and bone, or stable or poultry manure. An excellent ploy is to mulch within empty tyres around trees. This protects them initially from wind, rabbits, and drought. Thorn or thistle mulch in tyres discourages small browsers.

3. Gradually introduce poultry or light live-stock into the area, watching for damage.

4. Shift or add fences as the system proves itself, and continue to roll across the landscape.

5. Cull poorer plant specimens for pole timber, leaving selected high-yielding or strong trees and shrubs to continue growth.

■ ANIMAL ASSOCIATION AND INTERACTION

Like the rest of the system, animals are capable of beneficial and symbiotic interaction, as well as competitive, negative association. Design which takes advantages of these relationships comes from experience and observation, but some examples can be considered as below.

Poultry are scavengers and will salvage food that is wasted by other animals. On the other hand, chickens can pass on tuberculosis to cattle and thus, to humans. Pigs are also easily infected by chickens, so the two should not mix.

Cattle manure provides nutrients for pigs, which can follow cattle on pasture. On ground grains, four yearling steers will support one pig

163

through wastes alone. Ducks, also scavengers, will follow pigs, often gaining tidbits where pigs have been rooting.

Cats are totally destructive to small animal life (birds, lizards, frogs, etc.) so are a definite disadvantage. The insect pests of suburbia would be greatly reduced by frogs and lizards if cats were removed.

The succession of grazing species and their mixture must be regulated by considerations of disease transmission between species as well as by specific pasture conditions.

7.7
AQUACULTURE AND WETLANDS

A pond or lake can act as a mirror, a heat store, a run-off area, a cleanser of pollutants, a transport system, a fire barrier, a recreational asset, an energy storage, or part of an irrigation system. All this, and it is naturally productive as well.

Pond systems or aquacultures are far more productive and efficient than land-based systems due to a constant supply of water, nutrients in an easily-assimilated form, and a variety of plants and animals that can be eaten or sold. A mixture of fish, crayfish, molluscs, waterfowl, water plants, edge plants, and even land animals penned nearby takes advantage of different niches and foods in the system.

Most books treat aquaculture as *fish* culture, but there are as many useful plants as fish to be planted in water, and a great many algae, molluscs, even edible insects and frogs to be considered. We can design the system to make our main crops any of these: fish, water chestnut, wild rice, honey from marsh tupelo, bait fish, brine shrimp, freshwater snails, aquarium fish, water lilies as flowers or root sets, prawns, fish eggs, rushes or willows for basketry, fungi grown on rotting logs, and so on. All are "aquacultures". It is better to supply a small reliable specialist market—to grow red algae for carotene, for example, than to enter a mass market of pellet-fed trout or other highly capitalised ventures.

This chapter can only give some ideas for small-scale farm dams or home pond production. It is important to recognise that the more intensively farmed the system is, the more it will need research, careful planning, and sound management.

■ POND CONSTRUCTION
When planning and constructing aquaculture ponds, care should be taken to incorporate island refuges for breeding waterfowl, shallow shelves on the inside edges of the pond for

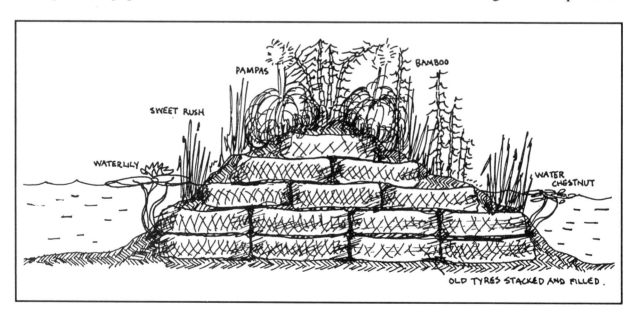

FIGURE 7.19 Tyre island for freshwater pond.

164

waterfowl forage plants, and a deep sump refuge for fish in areas where the dam is less than 3 metres deep and where summer temperatures are high. In addition, underwater refuges such as old tyres, earthenware pipes, and hollow logs protect smaller specimens from predator fish and cormorants.

Stabilisation of pond banks are by stepped log, tyre, or hand-cut planted ledges, using bamboo, pampas grass or other shallow matted-root species. Although shrubs can be planted, the root structure of large trees may eventually damage the bank and should be avoided.

When constructing a new pond or dam for fish aquaculture, do not stock with fish immediately. New dams do not have the ability of well-established dams to provide a range of natural foods. After the dam has filled for the first time, lay 5-10cm of straw around the waterline and trample it into wet soil. This not only minimises soil erosion but also provides cover and a food source for small aquatic insects. Water plants such as lily, cumbungi, water chestnut and even a small number of aquatic weeds (ribbon weed, water milfoil) also help to get the process started.

New dams are sometimes very muddy, and may need an application of gypsum (added at the rate of 560 kg/ha). Reduce the amount of silt coming into the dam from the inlet by grassing over the diversion drain or slope immediately above the dam. Careful management of the catchment area (planting vegetation, directing waterflow) is critical if the pond is not to be filled up with silt.

An island is built into a new dam simply by pushing clay into a large pile and topping with soil; alternatively, tyres are laid in a pile and filled with earth (**Figure 7.19**).

Stock should be fenced out of aquaculture dams; they will muddy the water, destroy vegetation, and may cause serious erosion problems.

■ POND DEPTH AND SHAPE

The number of fish that can be stocked in a pond directly relates to its *surface area*, not its depth or volume. Surface area controls the amount of food supply in and around the water. However, depth is also important in that fish must be able to escape to the pond bottom to cool off in hot weather and to avoid cormorants and other fish-eating birds. A usual figure is 2 to 2.5 metres deep. The following pond configurations are commonly-used throughout the world:

Ponds in Series: Fish of different age groups can be sequenced downstream, in a conveyer belt fashion (**Figure 7.20a**) In this way, food is supplied to the fish via a "trophic ladder" of nearby shallow ponds and marshes, which supply an overflow of live food to the main ponds, but which are safely isolated from predation so that fast-breeding food organisms can freely breed. As food is 70-90% of costs, it is far cheaper to breed it than to buy it.

Such an arrangement has the disadvantage that any parasite, disease, or water pollutant flows to every pond; although this is not usual in small operations, it must be an assessed risk.

Ponds in Parallel: the advantages are that each pond can be isolated for disease, and here again a food-species pond can lie above each production pond (**Figure 7.20b**). Note that "food species" may themselves be chosen to be either directly edible, or of use as bait fish. In general, ponds in parallel are more easily controlled, drained and serviced than ponds in series.

Canalised ponds: These are specifically suited to fish which depend for their food on pondside vegetation (grass carp, *Tilapia*) or land food (trout). Some of the most productive fisheries known are those of slow-flowing canals with ample food along them (some Swiss hill farms for trout are virtually contour canals on quite steep clay slopes). It is often easier netting fish in canals than in large, unshaped ponds (**Figure 7.20c**).

The ideal pond location and shape might be that of canals built through a marsh where food species breed, so that the canals are 20-30% of the total marsh area. The canals are stocked with predatory fish, which range through the marsh foraging on crustaceans and smaller fish. Netting fish for harvest occurs when the adjacent marsh area is low, say in dry summer seasons.

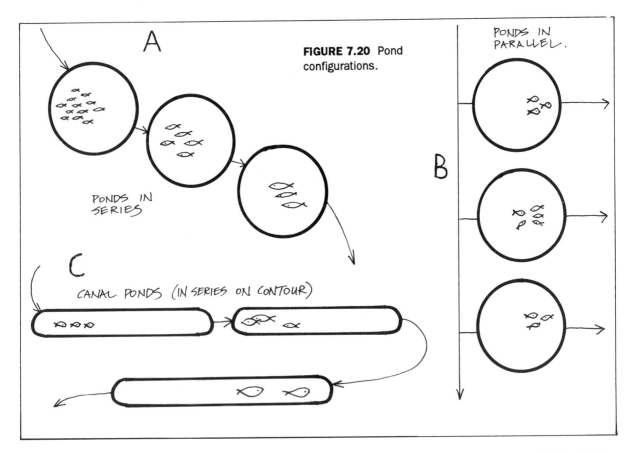

FIGURE 7.20 Pond configurations.

A PONDS IN SERIES

B PONDS IN PARALLEL.

C CANAL PONDS (IN SERIES ON CONTOUR)

POND SIZE

We need not think pond cultures are suited only to the standard half-acre pond; here are some useful products from small to large ponds:

• 1-2 square metres: Domestic watercress, taro, water chestnut, and a few frogs for garden pest control. A rare waterlily, or a small breeding population of a rare fish or aquarium plant.

• 5-50 sq. metres: A large range of plant foods, and at the upper pond size, enough carefully-selected fish for a family.

• 50-200 sq. metres: Commercial specialist crop, breeding stock, high value plants, and a full protein supply for a family. Supports a duck flock.

• 200-2000 sq. metres and up: Commercial for high value fish and shellfish. Larger size allows recreational uses.

(Note that every increase in size includes all the uses of the lesser sizes).

■ BENEFICIAL POLYCULTURE OR GUILD

Although an aquaculture system should be designed around a primary purpose (a particular fish, crustacean, or water plant), it is important to combine a range of beneficial aquatic species to fill all available pond niches, or to assist the primary product. The broad classes of aquatic organisms are these:

• Plants, from edge shrubs to fully submerged vegetation and phytoplankton.

• Invertebrates, both micro-organisms and shellfish or crayfish.

• Fish, from forage fish to plant, mollusc, and predatory species; up to 6 carefully-selected fish species might profitably occupy one pond and increase yield.

• Waterfowl, especially ducks and geese, and even pigeons housed over the pond.

Plants associated with ponds are:

• Edible root species, such as taro, water lily, lotus, and Indian water chestnut grown underwater in the banks or bottom, perhaps surrounded by an old tyre to mark their location.

• Floating aquatics such as Chinese water chestnut, kang kong, watercress, and the carpeting aquatics *Azolla* and duckweed. These may cover whole ponds, but can be raked up and either fed to animals (ducks thrive on them) or used as mulch on gardens or surrounding pond plants.

• Shallow edge plants of tall rush, cumbungi (cattail), or wild rice as frog and bird refuges.

• Seepage edge plants such as bamboo, papaya, banana, comfrey, elderberry and a short ground cover of grass or *Desmodium* (a creeping type). This ground cover keeps the banks stable and green, and is a source of forage for ducks and geese.

For aquatic animals, a range of specific different-level feeders are useful. Pond bottom feeders are those filtering or eating detritus and zooplankton, while top feeders are herbivores grazing algae and grasses. Ranging through the system are mid-level predators.

Detritus feeders are freshwater mussels and clams that live in the mud at the bottom of the pond. They can filter up to 900 litres a day of impure water through their systems and eject concentrated solutions (usually phosphorus) into the mud, which can then be used as fertiliser in orchards or crops when ponds are drained.

Other bottom feeders (on plankton) are crustaceans such as shrimp, crabs, and yabbies.

Herbivorous fish are those such as grass carp, which may completely clean up the pond of weeds and fringing vegetation. They are a fast-growing fish and reach market size in 3 months with an adequate supply of food. In Hawaii ponds are stocked with freshwater prawns as a main crop, with a secondary yield of grass carp eating fringing kikuyu grass. Ducks provide nutrient to the pond (ducks and fish are an excellent high-yielding combination).

Predator fish (e.g. bass, trout) are those which feed on other fish, and in a complex polyculture are screened off from the rest of the pond. Small fish and crustaceans enter the screened-in area and are eaten.

Such screened-off areas can be used for:

• Feeding and emergency aeration of water; this is used for eels, for example, saving energy on those few summer nights when aeration of the whole pond would be expensive.

• In the smaller ponds, high-value but *predaceous* fish can be kept to eliminate undersize fish from the larger ponds, via a screened separation which allows stunted or crowded fish to enter their area.

• Or the smaller pond sections can breed shrimp or minnow for larger fish in the main pond. Swingle (see references at the end of this chapter) estimates that 30% of any pond can be profitably screened off for forage fish and shrimp; nutrients are added to this part of the pond where shrimp rapidly take them up.

■ WATER QUALITY AND POND FERTILISATION

When building up a guild of species for a pond, prime considerations are to provide manures (fertiliser) for the pond system; to provide food for other organisms; to moderate the pond climate (edge vegetation); and to improve water quality, especially in the matters of waste utilisation and full use of foods.

Good quality water with a pH of 7-8 is best. If the water is too acid, nutrients in the soil are bound up and are not released into the water. It is common for pond bottoms to eventually acidify, and although lime may be added on the pond surface, the pond can be drained every few years. Many South East Asian farmers grow crops in duck-fertilised pond bottoms, then fill again for another fish cycle after liming. Dry-cycle crops can be grown in ponds every 2-4 years to take advantage of the generally high nutrient level built up in bottom muds for a high-value catch crop of melons, or a "luxury" grain such as wild rice.

Pond fertilisation is a key factor in raising yields, and can come from land animals, falling leaves, and other vegetation. Manures added to ponds increase plant growth and zooplankton blooms, which in turn increase the available food. Waterfowl on the pond, herbivorous fish feeding the edge, and land animals housed over the pond or along a canal leading to it all

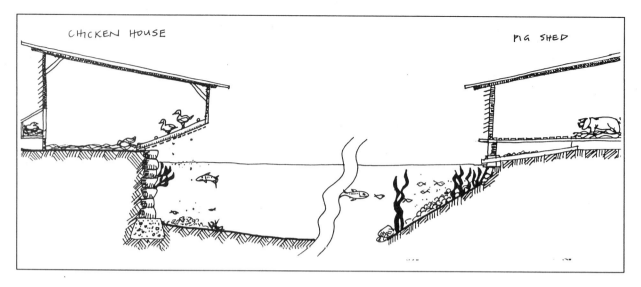

FIGURE 7.21 Widely used in Asia, animal manures are used to fertilise ponds. This is easily done if pens are located above the pond. Floors must be slatted so that manures fall into the water.

contribute valuable manures to the water (**Figure 7.21**). Shrimp, in particular, rapidly utilise manures from other species, and prawns fed on grass carp manure grow as well as on chicken manure, as they eat algae or diatoms produced on the surface of the manure.

Floating water plants (confined in ring rafts) and fringing rush beds help to remove or recycle pond nutrients to land crops by using them as mulch or compost. After fish are netted out, the nutrient-rich pond water may be used to trickle-irrigate land crop, with the result of a double production of leaf or fruit.

In densely-stocked or nutrient-rich ponds, the pond must be aerated in hot weather or fish will die. Paddle-pumps are usually used for aeration in commercial ponds, which are carefully monitored during critical times. However, in farm dams it is best to select species or to stock the dam so that aeration is not necessary. The height and shape of nearby tree species can also provide shade in hot weather; such shading by winter-deciduous poplars or willows may save aeration costs and provide leaf for worm beds.

Water quality and waste (fish and other faeces) removal is best achieved by including a set of scavengers, in particular freshwater mussels and surface algae eaters (*Vivipara* spp.), but also mud carp, catfish, and shrimp.

■ FEEDING FISH

Ponds should be designed as self-forage systems to minimise work. Food can be indirectly-provided through manuring the pond via ducks, planting fringing vegetation on which insects feed, e.g. silkworm larvae feeding on mulberry leaves are shaken into the pond occasionally, and placing insect traps on the pond surface. Planting flowering plants attractive to jewel beetles or wasps, and green "runner" ground covers such as *Tradescantia*, lucerne, comfrey, and other nutritious plant foods also aid fish feeding.

Direct feeding methods include raising high-protein worms and insects (cultured larvae) in special beds, or trapping insects in the garden or orchard for fish-food. We can breed or capture grasshoppers, fly larvae, pasture grubs, slaters (pill bugs), or even minnows, tadpoles and shrimp in smaller ponds. Rafts and ring nets in the pond itself can be added for special crop or food supply—worms and pill bugs breed as well on rafts as on land.

In addition to insect foods, high-carbohydrate grains, e.g. sorghum seeds, rice by-product, rice hulls, supplement protein foods. These are grown on site using nutrient-rich pond water.

■ STOCKING

Disease-free stock must be added to ponds from the beginning, so buy from a reputable dealer if possible.

Only as fish grow to optimum weight are natural food sources fully used; thus fast-growing minnows or shrimp can utilise these foods and store them in their bodies (as growth), for later use by predators.

As we increase the number of fish per unit area, the size of harvest decreases. Too few large fish or too many small fish show understocked or overstocked ponds respectively, with understocking the most common error on farms. The aim is not only to maximise yield, but to get a fish or plant of useful size. Over-mature fish and plants consume but do not grow at peak efficiency.

■ MARICULTURE

The same advantages of a mixed ecology of wildfowl, geese, fish, molluscs and algae apply to seawater or brackish ponds as they do to freshwater systems. The greatest advantage is a tide range of 1-9 metres, such as is found over most coasts. This range enables flushing and easy draining of ponds; the filling of higher-level impoundments for later release to lower ponds; and a flow of open sea species, fry, and algal forms as food.

Most shellfish and inshore species, including oysters, crayfish, eels, octopus, seagrasses, algae, shrimp, sand bivalves, and scale fish can be reared or managed in pond culture, raft culture, and fenced or impounded tidal areas. Many older civilisations, particularly the South Sea islanders, benefited from extensive, sophisticated fish traps, and today sea culture of oysters, mussels, and crayfish and lobster are a multi-million dollar industry.

Sea reef structures can be developed from tyres, broken or faulty earthenware, and stone to provide a substrate and shelter for larger forms of fish (octopus, crayfish). Lines of stone or woven fencing (long used in western Ireland) set out in shallow water "catch" algae as seaweed ponds.

Manurial input stimulates seagrass growth, and guano from seabirds, caught as liquid run-off from solid rafts or stony islands, provides the essential local phosphate and nitrogenous fertiliser for adjacent land crops. Even large artificial platforms have proved commercially viable off southwest Africa, where pelican and cormorant use these "islands" for roosting, and deposit tons of guano for fertiliser. In more humid climates, rain takes the guano into solution, so that storage tanks or covered solar evaporative pans need to be provided. Seagrass mulch and guano close the sea-land cycle of nutrients, and makes the growing of crops near seashores and waterways very profitable.

Some structures for mud-flats and intertidal sands are:
- reef walls of tyres, pipes, stone;
- drift fences to catch seagrass and direct fish;
- rafts for rope suspension of mollusc spat and algae, ring-rafts to rear fish in tideways (as in Ireland, where salmon are reared to adulthood in seaways);
- flow-governed tide pools to permit correct exposure for growing oysters;
- evaporative pans for salt, chemical, and brine-shrimp production (the latter as fry food);
- islands for marine wildfowl refuges and phosphate collection; and
- sub-surface (permeable) walls to retard tide flow in scoured estuaries.

Tide Traps

Where tides fall 1.2 metres or more on rocky coasts, tide traps are made from well-packed stone, so that rocky clefts or tidal flats are enclosed by a 90 cm wall (**Figure 7.22**). At high tide, the small school fish (garfish, squid, mullet, perch) enter over the wall, stay to feed on algae in the enclosed area, or are baited by crushed mussels, and are trapped inside as the tide falls. These can be fished or netted out, and used either for food or for stocking manured and managed ponds. A door in the wall allows the whole system to remain open when not in use.

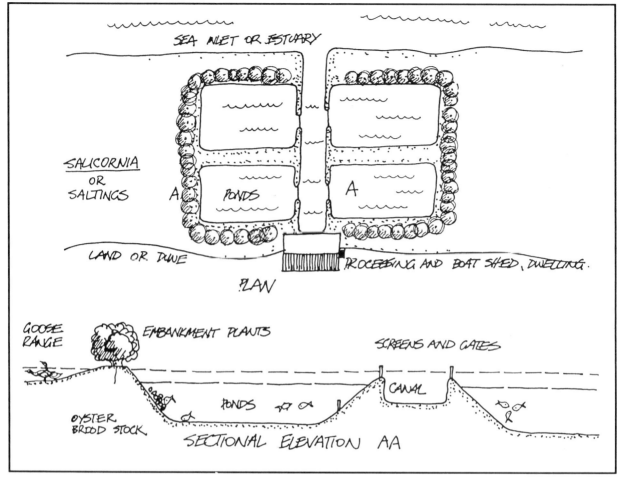

FIGURE 7.22 Plan and section of tidal pools.

7.8 REFERENCES AND FURTHER READING

Belanger, Jerome D., *The Homesteaders Handbook of Raising Small Livestock*, Rodale Press, Emmaus, 1974.

Chakroff, Marilyn, *Freshwater Fish Pond Culture and Management*, 1982, Peace Corps/VITA Publications No. 36E.

Fisheries and Wildlife Division (Victoria), various pamphlets including *Fish Farming in Farm Dams, Fish in Farm Dams, Fish Farming: Management of Water for Fish Production*.

Hill, D. and N. Edquist, Wildlife and Farm Dams, Fisheries and Wildlife Division and Soil Conservation Authority (no date given).

Maclean, J.L., *The Potential of Aquaculture in Australia*, Aust. Govt. Publishing, Canberra, ACT, 1975.

Reid, Rowan, and Geoff Wilson, *Agroforestry in Australia and New Zealand*, Goddard & Dobson, Box Hill, VIC 3630, Australia, 1986.

Swingle, H.S., *Biological Means of Increasing Productivity in Ponds*, 1966. FAO Symposium on warm-water pond fish culture 40-181, Rome, 18-20 May 1966.

Turner, Newman, Fertility Pastures and Cover Crops, Bargyla & Gylver Rateaver, Pauma Valley, California, 1977.

CHAPTER 8

Urban and Community Strategies

Before 1900 every city contained farms and orchards within the city. Although there are still such pockets of productivity left in the developing world, the modern need for more commercial buildings, industry, and living space has effectively pushed food-growing beyond the outskirts and into the distant countryside. Cities have become totally unable to provide for themselves in terms of food and energy, and now consume far more than they can produce.

Permaculture aims to bring food production back to urban areas, and to re-design or retrofit buildings to save and generate their own energy, using well-known energy-saving strategies and techniques of appropriate solar design for climate, weather-proofing, wind-power, trellis, insulation, low-cost transport, and cooperative power generation. It is only our passive dependence on city authorities that stops us from acting effectively. This chapter shows some of the ways in which urban and community self-reliance can be accomplished.

8.1
GROWING FOOD IN THE CITY

All cities have unused open land: vacant lots, parkland, industrial areas, roadsides, corners, lawns, areas front and back of houses, tubs, verandahs, concrete roofs, balconies, sun-facing glass walls and windows. Much of the current suburban vegetation is aesthetic rather than functional, and councils have small armies of people tending ornamental city plantings. It is only a matter of public persuasion and responsible decision to re-direct these activities to useful species, in a multi-dimensional and multi-faceted permaculture.

Parks, now largely open lawn, can be carpeted with edible and decorative understorey species such as blueberry, comfrey, currants, lavender, strawberries, etc. Useful pine-nut species can replace sterile cypress and pines, nuts replace eucalypts and barren hedgerow, and espaliered fruit can occupy walls and fences.

Urban woodlots grown around industrial zones and in greenbelts or undeveloped city land are not only aesthetically-pleasing but also filter pollutants from the air, produce oxygen, add to city fuel sources, and act as a wildlife habitat for birds and small animals. Some towns in West Germany now have urban forestry systems within and without the city boundaries. These provide firewood for sale to residents, woodchips and brushwood for composting, and a system of fast-growing trees for polewood and slow-growing trees for fine timber. With the addition of a mixture of easily-gathered food-producing trees such as oranges, apples, almonds, olives, pomegranates, dates, walnuts, etc. (chosen according to climate), city councils could reduce their dependence on rates collection or use these monies to fund recycling endeavours.

Leaves and clippings from urban permacultures are ideal compost and mulches for annual crops grown in intensive raised beds in backyards, or even on concrete patios and roof-tops (see Chapter 4 for urban gardening strategies).

Plants insulate against heat, noise and wind and give summer shade. Vines, moderators of summer heat, are a potential crop for warmer districts: scarlet runner beans, grapes, kiwifruit, choko, yellow and black passionfruit and hops are only some of the vines that can be used in this way.

Windows and greenhouses provide drying heat for long-storage products such as prunes, apricots, pears, apples and beans. Silvered insulation paper or mirrors will reflect light into dark corners. Walls can be painted black, or white, to act as heat radiators or reflectors.

The implications for energy conservation are obvious. Direct use of household produce means less use of expensive transport, packaging, and waste due to spoilage. Greater variety in diet and chemical-free food are an added bonus. The oldest and youngest can perform useful work in urban permacultural systems and the underemployed find useful activities in expanding the system. Much of what is now "garbage" can be returned to the soil, building up nutrients and lessening the waste production of the city.

8.2
PLANNED SUBURBAN AREAS (VILLAGE HOMES)

New suburban subdivisions can be planned for food production and energy self-reliance. Village Homes in Davis, California is such a development, and contains the following features:

• **Solar orientation**: every house faces the sun and incorporates passive or active solar space and water heating designs.

• **Water drainage**: all water run-off is led to swales, which provide a natural drainage system to replenish groundwater supples. Trees and shrubs are planted beside swales to take advantage of moist soils.

• **Greenbelts and common areas**: the space saved through the use of small front yards (fenced for privacy) and narrow streets is given over to community-owned greenbelts (for orchards, mini-parks, bike paths) and common areas. Houses are clustered into groups of eight, which have a say over the common area; they decide on the use and may plant vegetable gardens, develop a children's play area, or convert it to orchards, etc.

• **Shared resources and food production**: the community lands contain not only a meeting centre, playing fields, and swimming pool, but extensive areas for community gardens, grape orchards, and strip plantings of almonds, mandarins, pears, apples, persimmons, plums, and apricots. Twelve acres have been set aside for small-scale, non-commercial agricultural production; 50% of the development's total acreage will someday be in food production. In 1989, 60% of the residents' total food requirements was produced at Village Homes.

Davis itself is an energy- and water-saving city, with all new houses required to use solar energy and specific levels of insulation within the walls and ceiling. Street plantings of deciduous trees (shade in summer, sun in winter) are planted instead of evergreens. Drought-tolerant plants are required for public and commercial sites, and strongly encouraged for private yards. Shade trees are required by law in parking lots. Bike lanes and parking are especially catered for; 25% of all vehicle trips within Davis are now taken by bicycle.

8.3
COMMUNITY RECYCLING

A working example of town solid-waste recycling system is located in the Borough of

Devonport (Auckland, New Zealand). This innovative urban recycling scheme has been in use since 1977, when the rapidly overflowing tip (landfill system) was scheduled to be closed.

There are several key features that make the system work:

1. Separation of refuse at the source: Residents separate garbage into compostable materials, glass, paper, metals, etc., which means less time spent in sorting at the depot, and easily-available materials can be sold to recycling companies. The Council publicises the recycling scheme among residents, and hands out free calendars with collection times and dates each month.

There is a financial incentive to recycling: it is picked up free of cost. Unsorted refuse is picked up only if put out in special bags bought from the Borough (at a cost of $7 apiece!).

At the refuse tip itself, there are also separate bins for the following materials: scrap steel, hard plastics, tin cans, bottles, waste engine oil, paper, craft paper, and waste cloth. Firewood and reusable articles (such as furniture) are set aside for use by local residents.

2. Organic waste: The Council promotes the use of home composting to handle small units of domestic waste. It prepares publicity materials and home-made composting bins, and sells four types of bins at cost to residents. This means individual gardens receive the benefit, rather than concentrating the compost at the tip site.

For tree prunings and other compostable material, a large-scale composting operation is mounted at the depot. The material is chopped and shredded, and some animal manure is added to activate the heap; it is then formed into large windrows by a small bulldozer, and when finished, sold to local residents.

There is also a large garden at the site, made from compost, which produces vegetables for local sale. Trees and shrubs have been planted along the tip site, so that it has a pleasant appearance from the road.

3. Recoverable material: This includes scrap metal, tins, bottles, and newspaper; a contractor picks up this material at the same time as general refuse. The Auckland area offers a wide range of re-processing industries, so that the Devonport

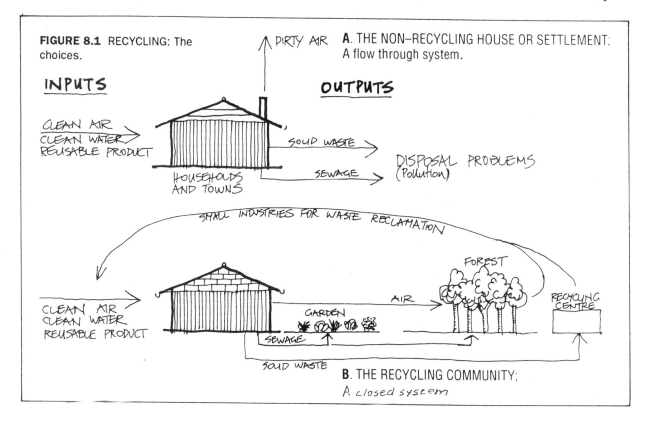

FIGURE 8.1 RECYCLING: The choices.

INPUTS

OUTPUTS

A. THE NON–RECYCLING HOUSE OR SETTLEMENT: A flow through system.

DIRTY AIR

CLEAN AIR
CLEAN WATER
REUSABLE PRODUCT

HOUSEHOLDS AND TOWNS

SOLID WASTE
SEWAGE

DISPOSAL PROBLEMS (Pollution)

SMALL INDUSTRIES FOR WASTE RECLAMATION

CLEAN AIR
CLEAN WATER
REUSABLE PRODUCT

GARDEN
SEWAGE
SOLID WASTE

AIR

FOREST

RECYCLING CENTRE

B. THE RECYCLING COMMUNITY: A closed system

Council has been able to sell most of its recovered materials.

Such an example shows that councils have no excuses for not recycling; not only does waste cost the ratepayers money, but there is also a vast waste disposal problem. It is up to ratepayers to elect officials who will recycle sewage and solid wastes, and to vote out waste-promoting councils, who "cost the earth".

Figure 8.1 shows the choices of recycling versus non-recycling.

8.4
COMMUNITY LAND ACCESS

Urban people who do not have access to land often work with others to grow food. There are many examples of this sort of cooperation all over the world. Some of the more successful are as follows.

■ COMMUNITY GARDENS

Community gardening is well-known in both urban and suburban areas. Residents clear rubble, put in water taps, build planter boxes, or whatever needs to be done to create garden space. They share water facilities, but generally have their own tools and garden allotments. To get such a project going, community interest needs to be stimulated, and signatures gathered to petition the local council. Council is lobbied to release vacant land within the city or town. A long lease is essential, which encourages residents to support and use the gardens without fear of abrupt change.

■ FARM-LINK: PRODUCER-CONSUMER COOPERATIVE

This is appropriate to high-rise or rental accommodation in a purely urban area, and was first developed in Japan. From 20 to 50 families link to a farm in the nearby countryside, usually with an already-established market gardener. Quarterly meetings are held between both parties to work out a wide range of products, from eggs to fresh produce to meats, with the consumers agreeing to accept all that is produced and to distribute the produce among themselves. Lower prices reflect this stable market with no packaging costs to the farmer.

As the "link" grows, the system might also accommodate holidays on the farm, educational workshops, and city help on the farm at peak work periods (planting and harvesting).

■ FARM CLUB

Garden or Farm Clubs suit families with some capital to invest as shares, with an annual membership. A farm is purchased by the club near the urban area (within 1-2 driving hours). The property is designed to serve the interests of members, whether for garden, main crop, fuelwood, fishing, recreation, camping, commercial growing, or all of these. People either lease small areas, or appoint a manager, depending upon the aims of the group and its finances. A management committee plans for the whole area (access, water, fences, rates, etc.), although individual projects/gardens/cabins may proliferate.

■ CITIES AS FARMS

There are several ways to use cities as farms. A community group or individual might collect surplus citrus and nut crop from trees around the city, at the same time distributing more trees to gardeners on contract for later product off the trees. Non-profit groups often collect unwanted produce from orchards, canneries, etc. and distribute them to the poor, or sell at a small profit to keep running costs down. This is known as a "gleaning system"; many thousands of tons of unwanted food are thus redistributed in the USA. Farmers or manufacturers take a tax deduction on gifts to a gleaning trust (any church or public trust).

Some city councils (Germany) carry on an active city forestry along roads and on reserves. From 60-80% of the city income is thus derived from city forest products.

CITY FARMS

A local group of 100 or more families forms a city farm association, and lobbies local or state authorities to allot from 1-80 hectares (preferably with a building) to a city farm. Again, a long-term, legally-binding lease is essential. Each city farm has a small management group and numerous volunteers. There could even be a few paid staff (for continuity purposes). On this land, the following activities are carried out (almost all are income-producing):

• Community garden allotments (if space allows) and demonstration gardens.

• Domestic animals (rabbits, pigeons, poultry, sheep, goats, cows, pigs, horses) for demonstration and breeding stock. Children are often involved in caring for animals after school.

• Recycling centre for equipment and used building materials such as bricks, pavers, windows and doors, aluminium and glass.

• "Gleaning" operations of surplus backyard, street and market garden produce. This is collected, sorted, and retailed. Herbs and other surplus from the demonstration garden can also be sold.

• Plant nursery of multi-functional plants: vegetables, groundcovers, shrubs, trees.

• Children and adult activities: seminars, demonstrations, training programs, educational outreach to develop community skills.

• Retail sales of seeds, books, plants and tools.

• Technical teams to provide home energy investigation and fitting of homes with weatherstripping for doors and windows.

• Information centre on food preparation, insect control, nutrition, home energy topics, etc.

The essentials of a successful city farm are that it lies in an area of real need (poor neighbourhoods), that it has a large local membership, and that it offers a wide range of social services to the area. Many city farms become totally self-supporting from sales of goods or services, plus modest membership fees. Government grants are sometimes needed in the first few years of setting up.

8.5
COMMUNITY ECONOMICS

Money is to the social fabric as water is to landscape. It is the agent of transport, the shaper and mover of trade. Like water, it is not the total amount of money entering a community which counts; it is the number of uses or duties to which money can be diverted, and the number of cycles of use, that brings financial independence to a community. We are talking about the linkages between the community and its finances, its base resources, and its legal structures. If you put a trading bank in a community that deals only with taking away basic resources, then what you've got is a pump that takes away the community's living and puts it somewhere else.

The following approaches, which have often been developed and applied by poor, depressed and often "powerless" groups, may be of use in your own community.

LETSYSTEM

LETS (Local Employment Trading System) centres in a community; every joining member must be willing to consider trading in local "green" dollars. Green dollars are "earned" by goods or services to others, and "spent" by using goods and services. Unlike a simple barter system (where two people trade only with each), a member who has credit can interact with any of the members in the LETSystem and can spend over the whole range of services or goods offered.

Green dollars are usually charged for labour, while federal currency is charged for the cost of the product or service, e.g. materials, petrol to and from the job, etc. Price is agreed upon by the individuals involved, and reported to the LETS Centre by the consumer. Anyone who wants work can offer services; they need not wait for "jobs". As only members can trade with each other, the community account is at all times balanced. An ideal member has many transactions, and accumulates modest debits and credits.

The currency, although equivalent to legal tender, is not issued and cannot be cashed in; it is kept only as a record of debits and credits. Any member can know the balance of any other member, and every member gets periodic statements of accounts. Any taxes applicable are the responsibility of members.

Anyone can start a LETSystem in their own community. See Resources section at the end of this chapter for addresses in both Australia and the USA.

■ REVOLVING LOAN FUNDS

These are community savings and loans associations appropriate for reducing community and household costs, and freeing more capital into the community. It is an easy matter to research what is lacking in the community, e.g. does the area make its own bread, yoghurt, sausages, shoes, clothes, and pots? Does it provide a wide range of services from haircutting to legal advice? If not, jobs are open and funds to start them can be available. Two successful examples are the SHARE and CELT systems of loans to community-based groups and businesses.

SHARE stands for Self Help Association for a Regional Economy. It is a local nonprofit corporation formed to encourage small businesses producing necessary goods and services for the region (in this case, the Berkshire area in Massachusetts, USA), which works in conjunction with a local bank in the area. SHARE members open a SHARE Joint Account with the bank. They receive only a small amount of interest (but this means small loans can be given out at less interest). The person receiving the loan must first collect references from people who know them as responsible and conscientious. They must show that the proposed business will attract customers from the community or even from outside the area. By doing this preliminary work, the borrower gets to know many people, and the community has a keen interest in seeing that the business succeeds.

CELT stands for Community Enterprise Loans Trust, a New Zealand-wide charitable trust to promote and support small businesses and cooperatives. CELT provides advice, runs training sessions, and gives out loans. CELT is funded by subscriptions from the public, by donations, and by government special schemes. Education and other work is funded by the interest from deposits and loans. The borrowing criteria is that the entrepreneur must be willing to work closely and regularly with CELT during the loan so that a business has the greatest chance to succeed. Jill Jordan of the Maleny Credit Union in Queensland reports that 85% of small businesses fail within their first two years of operation. In Maleny, however, businesses funded by the credit union and supported by the local community have a failure rate of less than 20%.

8.6
ETHICAL INVESTMENT

The past few years has seen a new movement towards innovative and consciously ethical financial systems. The rise of a large, popular, efficient set of services to divert public money towards beneficial ends is a reaction to the current misuse of money by governments, large aid agencies, banking institutions, and investment brokerages whose sole motive is profit or power.

We must not lend our money or effort to armaments, biocides, or to anything that will harm us or our environment. Instead of investing in our own destruction, we need to start directing our surplus monies into positive and life-enhancing projects.

The large amount of investment capital redirected through ethical brokerage firms in the U.S. and Australia is the tip of an iceberg which involves many thousands of ordinary people. They are members of guarantee circles, ethical credit unions, community loans trusts, common fund agencies for bioregions, or nonformal systems of labour and workday exchanges, barter systems, direct market systems, or no-interest, "green dollar" systems.

Moreover, existing banks, credit unions, co-operatives, and businesses are discussing the rewriting of their charters to include the values of earth care, people care, and the production of socially useful (or environmentally sensitive) products.

In earlier years, a negative ("non-buy") emphasis involved taking investments out of companies which polluted the earth and caused death through the manufacture of poisons, biocides, armaments, and other dangerous materials. As the ethical investment movement matures, however, this negative approach is evolving into a very positive search for, and willingness to fund and support, enterprises which:

• Assist conservation and reduce waste or energy use.

• Grow clean food free of biocides or dangerous levels of contaminants.

• Are involved in community reafforestation.

• Build energy conserving houses or villages.

• Produce clean transport or energy systems.

• Found cooperatives, self-employment ventures, or profit-sharing systems.

• Produce durable, sound, useful and necessary products.

Thus, local funds can establish small or large enterprises that are needed in the region, using money raised by residents. Brokers or enterprise trusts can direct surplus investment to socially and environmentally responsible industries and developments such as new, well-designed villages.

8.7
THE PERMACULTURE COMMUNITY

The global village community has been developing over the last decade. It is the most remarkable revolution in thought, values, and technology that has yet evolved. This book is intended to speed not the plough, but rather the philosophy of a new and diverse approach to land and living, and make the plough obsolete.

For myself, I see no other solution (political, economic) to the problems of mankind than the formation of small responsible communities involved in permaculture and appropriate technology. I believe that the days of centralised power are numbered, and that a re-tribalisation of society is an inevitable, if sometimes painful, process.

Unwilling as some of us are to act, we must find ways to do so for our own survival. Not all of us are, or need to be, farmers and gardeners. However, everyone has skills and strengths to offer and may form ecology parties or local action groups to change the politics of our local and state governments, to demand the use of public lands on behalf of landless people, and to join internationally to divert resources from waste and destruction to conservation and construction.

I believe we must change our philosophy before anything else changes. Change the philosophy of competition (which now pervades our educational system) to that of cooperation in free associations, change our material insecurity for a secure humanity, change the individual for the tribe, petrol for calories, and money for products.

But the greatest change we need to make is from consumption to production, even if on a small scale, in our own gardens. If only 10% of us do this, there is enough for everyone. Hence the futility of revolutionaries who have no gardens, who depend on the very system they attack, and who produce words and bullets, not food and shelter. It sometimes seems that we are caught, all of us on earth, in a conscious or unconscious conspiracy to keep ourselves helpless. And yet it is people who produce all the needs of other people, and together we can survive. We ourselves can cure all the famine, all the injustice, and all the stupidity of the world. We can do it by understanding the way natural systems work, by careful forestry and gardening, by contemplation and by taking care of the earth.

People who force nature force themselves. When we grow only wheat, we become dough. If we seek only money, we become brass; and if we stay in the childhood of team sports, we become a stuffed leather ball. Beware the monoculturalist, in religion, health, farm or factory. He is driven mad by boredom, and can create war and try to assert power, because he is in fact powerless.

To become a complete person, we must travel many paths, and to truly own anything we must first of all give it away. This is not a riddle. Only those who share their multiple and varied skills, true friendships, and a sense of community and knowledge of the earth know they are safe wherever they go.

There are plenty of fights and adventures to hand: the fight against cold, hunger, poverty, ignorance, overpopulation and greed; adventures in friendship, humanity, applied ecology, and sophisticated design—which would be a far better life than you may be living now, and which would mean a life for our children.

There is no other path for us than that of cooperative productivity and community responsibility. Take that path, and it will change your life in ways you cannot yet imagine.

8.8
RESOURCES AND REFERENCES

Morehouse, Ward, 1983, *Handbook of Tools for Community Economic Change*, ITDG Group of North America, PO Box 337, Crotonon Hudson, NY 10520 (Available from publisher). A basic explanation of land trusts, self-management, community banking, self-financing social investment, and SHARE programs. Highly recommended.

C.E.L.T. (Cooperative Enterprise Loan Trust): people's banking and seminars advisory services; includes S.C.O.R.E. (Service Corps of Retired Executives). P.O. Box 6855, Auckland, New Zealand.

LETS (Local Employment Trading System): organised credit/debit non-currency systems. Kits, games, software, information from: Michael Linton, Landsman Community Services Ltd., 375 Johnston Ave., Courtenay, B.C., Canada V9N 2Y2. **Or**, for Australia: Maleny and District Community Credit Union, 28 Maple St., Maleny QLD 4552, Australia.

MONEY MATTERS, Mezzanine Level, 27 - 31 Macquarie Place, Circular Quay, Sydney, NSW 2000, Australia. Ethical investment advisory service.

S.H.A.R.E. (Self-Help Association for a Regional Economy), P.O. Box 125, Great Barrington, MA 01230, USA.

Appendix A

LIST OF SOME USEFUL PERMACULTURE PLANTS

Most of the species below are perennial, although some annuals are included. This list is by no means complete; it is intended only as an informal start to your own local permaculture species lists. The plants below range from temperate to tropical climates; many temperate species can also be grown in the subtropics or highland tropics. In most cases heights are given (in metres - m) but these will vary according to climate, care, soils, and cultivars.

ACACIAS (*Acacia* spp.)
Leguminous trees and shrubs ranging from 3-25m, species growing from arid regions to the tropics; often spiny. USES: Some species are important fodder plants of drylands, with leaves, pods, and seeds used; firewood and (some species) timber. Nitrogen-fixing; Fukuoka planted silver wattle (*A. dealbata*) in his fields to boost production. Erosion control.

Fodder: Mulga (*Acacia aneura*) widespread in Australian drylands, fast-growing and palatable to stock; to 7m tall. Camel thorn (*Faidherbia albida*) thorny tree to 25m; foliage and pods important fodder yielding 135 kg pods/tree in Sudan. Deciduous in wet season, full leaf in dry. Myall (*A. pendula*) grows on heavy soils where no other trees will grow (protects soil and gives shade as well as fodder). Other fodder trees are *A. salicina* (native willow), *A. senegal*, *A. seyal*.

Timber: Blackwood (*Acacia melanoxylon*), fast-growing, long-lived, cool climate acacia used in fine furniture (in warm climates *A. melanoxylon* is a scraggly, short-lived tree). Silver wattle (*A. dealbata*) and hickory wattle (*A. falci-formis*) also important timber trees.

ALBIZIA (*Albizia lopantha, A. julibrissin*)
Leguminous, evergreen, quick-growing trees with feathery leaves. Height: 9-15m. Warm temperate to tropical climates.

USES: Shade tree, with ornamental leaves and flowers. Windbreak if lopped to encourage bushiness. Pioneer tree; in the tropics, chili peppers, pineapples, banana, and fruit trees are grown under widely-spaced albizia, providing a 3-tier productive system. Most species are palatable to stock (*A. lopantha, A. chinensis*). Nitrogen-fixing.

ALDER
(*Alnus* spp.): Fast-growing, short-lived trees mainly forming dense thickets. Height: 10-25m. Although not legumes, are nitrogen-fixing, and create a thick, black humus. Useful if already present for rough mulch, composting. Use as a nurse crop for other trees; provides shelter, mulch, and nitrogen. Can eventually be cut out altogether, or a few trees allowed to grow on for nitrogen-fixing, mulch. As firewood it may burn too hot, but stickwood is useful. Some *Alnus* spp. are *A. tenuifolia* (mountain alder), *A. crispa* (downy alder).

AMARANTH (*Amaranthus* spp.)
Upright annuals to 1m of which grain amaranth (*A. hypochondriacus*) and leaf amaranth (*A. gangeticus*) are most valuable. Grown in full sun or even partial shade; grain amaranth needs a 90-day growing season to set seed. Temperate areas through highland dry tropics.

USES: Grain amaranth a high protein crop (18%); seeds eaten popped or ground into flour. Leaves eaten raw or cooked. Leaf amaranth grown throughout year in warm climates; tasty leaves are bright red and green. Valuable vitamin and mineral plant. Chicken fodder (seeds); leaves for stock—can be turned into silage. Cover crop.

ARRACACHA (*Arracacha xanthorrhiza, A. esculenta*)
Also known as Peruvian parsnip. Grown in high-altitude tropics to subtropical climates. Herbaceous perennial, producing large, starchy roots. Propagated by tubers.

USES: Eaten like potatoes or cassava. Coarse main rootstocks and mature leaves fed to animals. Young stems for salads. Excellent understorey crop.

ASPARAGUS (*Asparagus officinalis*)
Perennial rootstock with new, edible shoots each year, yielding well for at least 20 years if manured and watered. Yields after 3 years, in spring. Easily propagated in winter by crown division. Naturalises along sandy watercourses, though does not produce large stalks as does manured asparagus. USES: human food, bank stabilisers for sandy steams. Temperate to subtropical climates.

AUTUMN OLIVE and **RUSSIAN OLIVE** (*Elaeagnus umbellata, E. angustifolia* & other spp.)
Fast-growing, nitrogen-fixing shrubs to 4.5m and 20m respectively; Autumn olive forms thickets or hedges when clipped. Tolerates poor soil, drought. Likes full sun, although other species will tolerate partial to full shade.

Temperate and cold area plants.

USES: Good windbreak and erosion control plant. Edible berries for birds and poultry; cold area chicken forage plant. Ornamental screen hedges. Silverberry (*E. commutata*) and cherry elaeagnus (*E. multiflora*) also important wildlife, poultry berry plants.

AZOLLA (*Azolla* spp.)

Free-floating, small water ferns (red or green) which contain a nitrogen-fixing bacteria (*Anabaena azollae*). All climates, although dies back in hot weather. USES: Duck fodder. Nitrogen mulch for rice or taro crop for nitrogen. Can be skimmed off surface of ponds and used as a rich mulch on adjacent crops; or ponds drained, *Azolla* turned under, and crops grown.

BAMBOO (1250 species)

Two main types are running bamboos and clumping bamboos. Generally the tropical/subtropical varieties are clumpers and the temperate varieties are runners. In the case of runner bamboos, care must be taken so that they do not become rampant; they do not cross water, so may be contained on an island in a dam.

Bamboos grow from the equator to about 40° north and south. Propagation is by division of clumps, rhizome cuttings, and basal cane cuttings; bamboo grows best in rich organic soil with plenty of water.

USES: human food (clumps are hilled to produce large, tender shoots) and foliage as animal forage (some species such as *Arundinaria racemosa*, *Sasa palmata*). Structural: stakes, fishpoles, spears (small canes), building frameworks, concrete reinforcing (big canes). Clumps: windbreak, steep bank stabilisers. Other: utensils, mulch, artisanry.

BLACK LOCUST (*Robinia pseudoacacia*)

Deciduous tree 10-20m, thin foliage, lives up to 200 years. Grows rapidly and forms thickets by root suckers (very aggressive). Very hardy and suited to cool areas, poor soils.

USES: Pasture improver on very poor country (nitrogen fixer); erosion control; windbreak tree; bee forage; seed for poultry; and timber suited to beams, tools and shafts. Poles last over 20 years untreated in the ground.

BLACKBERRY, RASPBERRY (*Rubus* spp.)

Cultivars include boysenberry, loganberry. Vigorously-growing prickly thickets (some thornless varieties have been developed). High-value commercial crop on trellis. Blackberry easily becomes rampant, spread by seeds and tip-rooting. Can be marooned on islands. Blackberry (*R. laciniatus*) has a thornless variety (Oregon thornless) which is best for gardens. Loganberry and boysenberry are preferred cultivars, with very large berries. May need netting against birds. Bee forage.

BLUEBERRY, HUCKLEBERRY, CRANBERRY (*Vaccinium* spp.)

Deciduous shrubs from 2.5cm to 3.6m tall; cool temperate to subtropical climates. Tolerate partial shade or full sun. USES: Understorey berry crop. Most species good bee forage.

High bush blueberry (*V. corymbosum*) grows to 1.2-3.6m and is grown as a commercial crop, needs to be netted against birds. Low bush blueberry (*V. angustifolium*) can be used as a groundcover (8-20cm); avoid frost pockets.

Huckleberry (*V. membranaceum*, *V. ovatum*) are not commercially grown for berries, but these are tasty for human use; also poultry forage. Evergreen huckleberry produces best in partial shade. Species grow 30cm-3m tall.

Cranberry (*V. oxycoccus*) is about 25cm tall; it is an evergreen, prostrate undershrub, and grows well in peat bogs, with soil pH of 3.2 to 4.5. A constant water supply is necessary for good fruiting, but plants should not be swamped. Rich humus and thick mulches are ideal. Avoid planting in known frost pockets; fruits must ripen before hard frosts. High-value commercial crop.

BORAGE (*Borago officinalis*)

An upright, self-seeding annual to 0.6m at maturity. Can be grown in full sun or partial shade; tolerates poor soils but needs regular watering. Easy to propagate in large quantities; sow seed in spring. Temperate climate. USES: Good bee forage, with a long flowering season. Leaves and flowers in salads. Compost/manure tea with comfrey: rich in potash and calcium; breaks down very quickly. Medicinal properties: anti-inflammatory.

BROAD BEAN (*Vicia faba*)

Annual legume 0.5-1m; temperate to subtropical climates, likes full sun but grows well in cloudy maritime climates over winter. USES: Human food—young leaves, pods, beans (fresh or dried). Also used as stock fodder. Cover crop over garden beds, fields; green manure crop and nitrogen fixer, with crop cut and used for mulch before flowering (nitrogen stays in the soil).

CAPE GOOSEBERRY (*Physalis peruviana*)

A perennial, tender, creeping bush of the tomato family (Solanaceae) with small greenish-yellow fruits surrounded by a papery calyx or husk. Fruits ripen in late summer and are used fresh or stewed. Used in Mexico as a hot sauce when mixed with chillies and onions. Easily frost-damaged; grown as an annual in cold temperate climates.

CAROB (*Ceratonia siliqua*)

A long-lived tree 5-15m grown for its sugary pods. A tree of the Mediterranean, it does best in dry temperate climates and can tolerate poor soil conditions. Frost damages flowers and young fruit, but not the trees; very wet weather in autumn can rot the ripening pods. A leguminous tree, although does not fix nitrogen.

USES: Human food: ground meal is a chocolate or coffee substitute, widely used in health food products. Pods as stock feed for energy and protein concentrate (ground as meal or fed whole to large animals). Yields in Mediterranean climates are 45-225 kg/tree. The seeds yield a gum with water-absorbing qualities, used in cosmetic and chemical industries.

CASSAVA (*Manihot esculenta*)
Lowland tropical crop, with starchy tubers. Widely used in Africa, South Pacific, Latin America. Grown on ridges or mounds, interplanted with annual food crops. Can withstand neglect, grows in nutrient-poor soils; tolerates drought (except after propagation). Can be kept in the ground until required.

USES: Eaten (after peeling) boiled or baked. Dried slices may be kept for several months; cassava flour is made by grinding these dried chips. Fermented pulp is eaten in West Africa. Starch, or tapioca, is used for puddings, biscuits, and confectionary.

CHESTNUT (*Castanea mollissima, C. sativa*)
Large, spreading deciduous tree to 30m; long-lived. Grafted trees yield in 7-9 years. Temperate Mediterranean climates; tolerate dry conditions. Like well-drained soils. Need cross-pollination for best results. May not bear well in climates with cool summers.

USES: As food: Spanish or sweet chestnut (*C. sativa*) important commercial crop in Europe, while Chinese chestnut (*C. mollissima*) grown in U.S. because of resistance to blight fungus. Chestnuts are eaten whole, roasted and husked, or ground for sweet flour, rich in starch. High-grade stock fodder, especially for pigs.

CHICORY (*Cichorium intybus*)
Herbaceous perennial long used as a vegetable in Europe and the Orient; grows from 0.6-1.6m. Likes full sun and grows from temperate to subtropical regions. Naturalises in fields and on disturbed soils.

USES: Bee forage; early and long-flowering. Roots roasted for coffee-like beverage. Mineral-rich leaves (from deep taproot mining the soil) excellent component in pasture as forage crop; improves milk quality and quantity. Medicinal (both human and animal); used for rheumatism, eczema, blood diseases.

CHINESE WATER CHESTNUT (*Eleocharis dulcis*)
Aquatic rush with edible culms, grown in shallows or damp mud. Subtropics/tropics: can be grown wherever there are 8 frost-free months. May need to be netted against ducks when green shoots are emerging. Caution: As with many aquatic plants, these may accumulate heavy metals, so make sure pond water is not polluted (or use these to help clean up water; do not harvest). USES: Valuable human food, high in carbohydrate, used extensively in Asia.

CITRUS (*Citrus* spp.)
Wide range of evergreen shrubs or trees to 10m, including lemon, lime, cumquat, orange, grapefruit, mandarin. Dry, warm temperate (Mediterranean) climates to tropics. In marginal temperate areas, place in warm, sunny position. Tree can withstand light frost, but frost at -2° C kills flowers and young fruit. Need shelter in high wind areas.

USES: Fresh fruit or juice, marmalade, concentrated for cordials. High vitamin C source, especially if white pith is also eaten. Waste pulp fed to cattle. Peel is source of essential oils (used in flavouring and perfumes); also provides pectin.

CHOKO or CHAYOTE (*Sechium edule*)
Herbaceous scrambler, vigorous, perennial on thick rootstock. Subtropics to tropics; not hardy to frost.

USES: Roots used for starch, boiled or baked; young shoots eaten as a salad, steamed. Most commonly-eaten is

181

CHOKO
(SECHIUM EDULE)

the fruit, a large bland vegetable which can be baked, steamed, or fried with other vegetables. Used to smother less vigorous plants such as lantana, and is a good roof covering for summer. Pig and poultry food.

COMFREY (*Symphytum officinale*)
Herbaceous perennial to 1m high. Dies down in winter, except in mild climates. Easily propagated by root division; any part of the root crown will grow. Clumps of comfrey will stay in one place, but if dug or rototilled will spread quickly. High yields on fertile, watered country. 20-25% crude protein.

USES: Excellent bee forage. Stock fodder if fed in limited quantities (overfeeding has been shown to cause some liver damage in animals). Medicinal herb: roots dried, powdered and used in ointments for bruises, arthritis, broken bones. Vegetable source of vitamin B12, and can be used sparingly in salads, cooking. Rich source of mulch (high potash) and is combined with other leaves and manures for a nutrient-rich "manure tea".

CURRANTS and GOOSEBERRIES (*Ribes* spp.)
Small deciduous shrubs (0.5-1m) tolerating partial-shade; good hardy understorey bush tolerating neglect. Hardwood cuttings taken in autumn root easily. Bear 10-20 years if properly cared for. Mostly temperate plants.

USES: Tasty small fruits which can be eaten raw or made into juice, wine, jelly. Wildlife forage food, including birds and poultry (plants may need to be netted if used entirely for human food). Edible species: black currant (*R. nigrum*), golden currant (*R. aureum*), red currant (*R. rubrum*) Excellent bee forage. Also ornamental, especially golden currant and red-flowering currant (*R. sanguineum*). Gooseberries (*R. grossularia*) grow successfully in rock crevices; like well-drained positions.

CUMBUNGI or CATTAIL (*Typha latifolia, T. orientalis*)
Dense, herbaceous perennial to 4m; grows in full sun or shade around pond edges. **Caution:** Can be an invasive weed. Temperate to subtropical climates.

USES: Shoots edible, used like asparagus. Roots are peeled, cooked or grated raw. Seeds, roasted, have nutty flavour. Animal forage, mainly roots, especially for pigs. Weaving material, basketry. Duck and water fowl habitat. Seed head is of downy material; can be used as tinder. Extracts pollutants from water.

DANDELION (*Taraxacum officinale*)
Small perennial herb with yellow flowers early spring to late autumn. Grows in temperate to subtropical areas and is a common weed on lawns, pastures. Grows in full sun or shade.

USES: Leaves, roots, flowers eaten; roots are used as a coffee substitute. Flowers can be used to make wine. Important bee forage with early and long flowering; high pollen yield. Forage crop, improves milk quality and quantity; good mix with lucerne.

DAYLILY (*Hemerocallis fulva*)
Herbaceous perennial to 0.6m, temperate to subtropical climates. Tolerates partial shade; useful understorey plant. USES: Edible shoots, flowerbuds, flowers, tubers. Low-maintenance plant; erosion control on hillsides. Ornamental. Grow under trees as part of guild with marigolds, dill, nasturtium, etc.

DUCKWEED (*Lemna minor*)
Perennial floating aquatic of ponds (likes quiet water); temperate climates. May completely cover a pond and exclude light. USES: Duck, goose, fish

forage; may have potential as chicken, pig food. Can be skimmed off ponds and used as high-nutrient mulch material. May take up heavy metals in polluted waters.

ELDERBERRY (*Sambucus nigra, S. canadensis*)
Deciduous shrub to 6m, temperate climate, tolerates full sun to partial shade. Easily propagated from cuttings.

USES: Hedgerow shrub; windbreak. Ripe berries make wine, dye, conserves (should not be eaten raw). Flowers fermented with lemon juice and peel as a beverage, or infused in hot water for respiratory inflammations. CAUTION: leaves, roots, stems and unripe fruit may be poisonous to humans and to stock.

FEIJOA (*Feijoa sellowiana*)
Also called pineapple guava, although not a true guava. Evergreen shrub 4-6m. Warm temperate areas to subtropics; grows in cool climates but fruits only in hot summers (place in sunny location). Needs shelter from wind. Grown commercially in New Zealand. If growing from seed, notice round tips on leaves in nursery beds; these indicate large-fruited forms and should be selected. Yields 3-4 years from cuttings (taken in summer).

USES: Fruit for dessert, conserves. Petals of flowers are very sweet and used in salads. Ornamental.

FENNEL (*Foeniculum vulgare, F. dulce*)
An upright self-seeding biennial or short-lived hardy perennial with umbel-shaped flowers in summer which attract beneficial insects (insectary plant). Grows in poor soils; naturalises along roadsides in temperate climates. Grows both in full sun or full shade.

USES: Seeds for culinary purposes; seeds and roots medicinally. Foliage as fresh herb, and root of Florence fennel (*F. dulce*) used in salads (crispy like celery, but with an anise flavour); prefers rich garden soil. Stock fodder in controlled quantities is medicinal. Suppresses grasses.

FIG (*Ficus carica*)
Deciduous shrub or tree to 8m; widespread in Mediterranean climates and marginal subtropics (not too wet). Likes full sun; will shade out anything planted underneath

unless pruned. Propagated by cuttings. Important commercial crop, eaten fresh or dried. Useful chicken and pig forage. Mulch from dead leaves in autumn.

FILBERT, HAZEL (*C. maxima, Corylus avellana*)
Many varieties, most producing edible nuts (filberts and hazelnuts). Small, deciduous trees or thicket-forming shrubs to 6m; long-lived to 150 years. Grafted varieties start yielding in 5-6 years, with peak nut production at 15 years. Major commercial production in dry, Mediterranean countries, but also suited to cool temperate. Needs cross-pollination. Tolerates shade, but for nut production needs sun; yields best on an edge. Well-drained, fertile soil is best.

USES: Nuts for human food; also as animal forage (low-grade or small nuts). Good hedgerow tree which can be coppiced for poles, stakes, etc.; may need wind shelter in first years.

GLIRICIDIA (*Gliricidia sepium, G. maculata*)
Fast-growing, vigorous deciduous tree to 9m; out-competes most tropical grasses. Grows in tropical and subtropical climates. Legume tree.

USES: Widely-used shade tree for banana, coffee, young cocoa. Can be topped to produce material for green manuring. Tolerates repeated coppicing and is used in alley farming and for firewood. Also useful as a firebreak and tropical bee fodder. Durable wood for poles, fenceposts, and stakes. Legume tree.

GUAVA (*Psidium guajava* & other spp.)
Shallow-rooted shrub or small tree, 3-10m; can produce suckers. Adaptable to wide range of soils; susceptible to frost. Drought-tolerant. Sometimes become rampant as seeds are carried by birds.

USES: Fruit eaten fresh, although its numerous seeds make it best for conserves, jellies, paste, juice. Very high vitamin C (2-5 times that of oranges). Strawberry guava (*P. littorale*) hardier, marginally suited to cool areas; place in warm, sunny position.

GINGER (*Zingiber officinale*)
Herbaceous perennial of the humid tropics and subtrop-

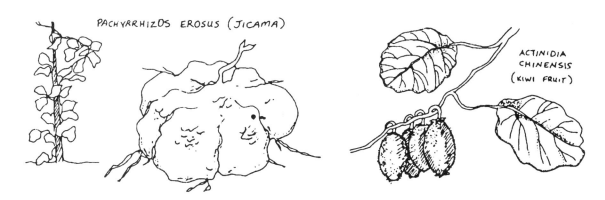

PACHYRRHIZOS EROSUS (JICAMA)

ACTINIDIA CHINENSIS (KIWI FRUIT)

ics; reaching to 90cm. Easily propagated by rhizomes. Often grown commercially as an intercrop with coconut, coffee, citrus, and turmeric (which provides partial shade to young ginger). Rhizomes eaten fresh or preserved for flavouring (candied, dried and powdered).

GRAPE (*Vitis vinifera* & spp.)
Long-lived, deciduous perennial vine, preferring some chill factor for fruiting, but many varieties and cultivars are adapted to a wide climatic and soil range. Planted on trellis, although in ancient times allowed to scramble on mulberry and fig trees.

USES: Fresh fruit; also dried (raisins), wine, juice. Young leaves are used as a food wrapping in cooking (Greek dolmas). Seeds are an excellent cooking oil. Deciduous vines to block summer sun from houses.

HAWTHORNS (*Crataegus* spp.)
Tough, thorny, deciduous shrubs/trees 2-7m high; slow-growing but long-lived (100-300 years). Tolerate partial shade, poor soils.

USES: Edible berries for jellies, conserves. Hedge and windbreak plant for temperate climates, grown extensively as hedgerows in England. Habitat for birds: shelter, nesting and food; useful for poultry. Good bee forage. Coppice wood. Black hawthorn (*C. douglasii*) produces best fruits for human consumption. English hawthorn (*C. monogyna*) makes a narrow, dense hedge. Popular southern European variety is Mediterranean medlar (*C. azarolus*).

HICKORY (*Carya ovata, C. laciniosa, C. ovata*)
Large, deciduous trees (18-45m) yielding nuts through winter to spring; form upright, cylindrical crowns. Yields often irregular, need cross-pollination. **PECAN** (*C. illinoensis*) most important nut tree of the genus. It needs 150-200 frost-free days, without extremes of cold or heat; suitable for subtropics but grown even in New Zealand.

USES: Nuts as human food; inferior nuts as forage for pigs (also for chickens if cracked and soaked). Excellent wood for tool handles (very tough) and charcoal (imparts flavour to hams in smoking process).

HONEY LOCUST (*Gleditsia triacanthos*)
Deciduous tree 6-40m; very thorny when young, although thornless cultivars have been developed (*G. triacanthos inermis*). Trees have open canopy which allows clovers and pasture to be grown underneath. Frost- and drought-hardy; likes temperate regime of hot summers, cold winters. Tolerates most soils. Although a legume tree, nitrogen-fixing nodules have not been observed in the roots.
Yields up to 110 kgs of pods per tree at years 8-9; at 86 trees/hectare, pod production equivalent to 10 tons/hectare of oat crop. Transplants easily, grows in full sun. Seed pods need to be gathered from trees as soon as they fall in mid-autumn and the seed scarified or boiling water poured over them (and soaked). Select high-yielding, thornless varieties.

USES: Pods are high in sugar (27-30%); pod and seeds 10% protein. Excellent stock fodder, ground or whole, especially during drought or at the end of summer pasture. Durable, quality timber. Excellent bee forage. High sugar content means potential for fuel production, molasses, wine.

HOPS (*Humulus lupus*)
Long-lived (80-100 years) herbaceous perennial climber. Propagate from root cuttings. Naturalises on swamp edges and river banks, scrambles in shrubs and trees or can be wound on hanging cords, wires.

USES: Mainly grown for beer flavouring, but also used as a pillow filling and mild narcotic (hops steeped in sherry to enhance calm and sleepiness). Shoots and tips used as steamed green. Browsed by sheep, geese when young, although sheep can be used in plantations from late spring to winter to browse the grass beneath the hops as commercial hop growers often cut vines to the roots.

HORSERADISH (*Armoracia rusticana*)
Herbaceous perennial 0.5-1m growing from large, edible root. Grows best in cool climates; likes full sun but can grow well in partial shade and useful as an understorey plant. Propagate by root division; all the pieces grow (like comfrey). Root eaten as a condiment. Medicinal uses are as a diuretic, for infections and lung problems.

ICE CREAM BEAN (*Inga edulis*)
Medium, leguminous tree to 12m; subtropical and tropical climates. White fruit pulp from pods used in desserts (said to taste like ice cream). Shade tree for coffee and tea plantations; mid-level understorey tree. Nitrogen-fixer.

JERUSALEM ARTICHOKE (see SUNROOT).

JUJUBE (*Ziziphus jujuba*)
Also called Chinese date. Deciduous tree to 12m; sometimes a large, spiny, dense shrub. Thrives in hot dry regions, alkaline soils, and can withstand severe heat, drought, and some frost. Propagation by stratified seed or root cuttings.

USES: Fruit can be eaten fresh, dried, pickled (resembles dates). Leaves and fruit useful fodder for stock, pigs. Trees coppice well and produce good firewood. Leaves used to feed the tasser silkworm.

KANG KONG (*Ipomoea aquatica*)
Aquatic floating herbaceous perennial found throughout the tropics. Young terminal shoots and leaves used as spinach; rich in minerals and vitamins. Vines are used as

fodder for cattle, pigs; also fish food. "In Malaya it is widely grown in fish ponds by the Chinese who feed it to their pigs; the pig manure is used to fertilise the fish ponds; thus fish, pork and spinach are provided." (*Tropical Crops - Dicotyledons*, J.W. Purseglove, 1968).

KIWIFRUIT (*Actinidia chinensis*)
Also called Chinese gooseberry. Large, woody, deciduous climber, trellised at 2.5m, forming a bramble. Dioecious (male and female plants), although male and female may be grafted on one vine. Needs a strong trellis system. Tolerates frost; grown from temperate climates to subtropics. Needs shelter from wind. *Actinidia arguta* tolerates heavy frost; has smaller, astringent fruits, but hybridised with kiwifruit will produce sweeter yields.

USES: Delicious fruits; for eating, wine, conserves. May be fed to pigs and chickens if fruit set is abundant; also a high-value commercial crop. Useful deciduous shade vine for pergolas, patios.

KURRAJONG and BOTTLE TREE (*Brachychiton populneum* and *B. rupestre*)
Hot, dry climate fodder trees suited to agroforestry. Large trees. Have deep tap-roots; do not compete with crops or pastures. Can be easily coppiced. USES: Leaf fodder, especially as drought rations for sheep and cattle. Leaves lack phosphorus, which must be provided by stock licks. Bottle trees (*B. repestre*) are often cut down completely to feed soft inner pith to cattle in extreme drought; these must be replanted.

LAB-LAB BEAN (*Lab-Lab purpureus*—Syn. *Dolichos lab-lab*)
Herbaceous perennial legume, often grown as an annual; 1.5-6m tall. Subtropical to tropical evergreen or summer herbaceous climber. May become rampant, but managed by slashing 3-4 times a year or grazing by sheep, goats or cows. In subtropics dies back in light frost and can therefore be interplanted with grains. Tropics: remains green in dry season.

USES: Young leaves eaten raw or cooked, ripe seeds as split peas, or sprouted, boiled and mashed to a paste, then fried. High biomass forage crop (either green or as hay or silage) A useful dryland trellis crop for a sun shield (must be watered). Excellent green manure plant and cover crop; cut and use as mulch. Often grown in rotation with commercial crop to provide nitrogen.

LAVENDER (*Lavandula vera, L. dentata*)
Small, woody shrub easily grown from cuttings. Suited to cool areas and is drought-resistant (originally a Mediterranean mountain plant). Well-drained, alkaline soil is best.

USES: Ornamental hedge plant, creating an "edge" in gardens; excellent bee forage. Flowers and leaves used medicinally. Lavender oil is a powerful germicide and insect repellant; dried flowers keep moths out of stored linen and clothes. Place sachets of lavender flowers in clothes drawers.

LEMONGRASS (*Cymbopogon citratus*)
Perennial medium-sized "grass" of subtropics and tropics. USES: Lemongrass tea, and flavouring used in Asian cooking. Excellent border plant in gardens and orchards to create edge; cut and use for mulch. Erosion control on slopes when planted in rows along the contour; will catch and hold silt.

LESPEDEZA (*Lespedeza* spp.)
Sericea (*L. cuneata*) is a perennial legume (similar to clover) common in temperate zones. High-value animal fodder, hay and soil improver (nitrogen-fixer). Also used

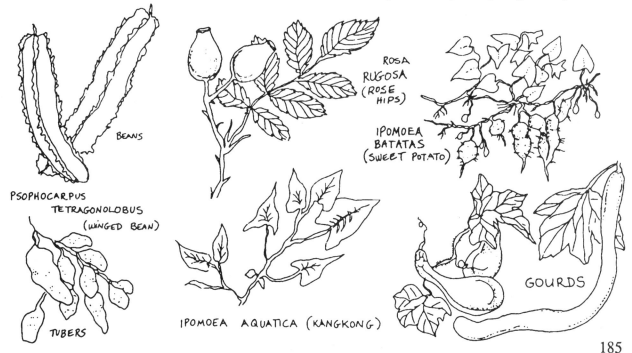

PSOPHOCARPUS TETRAGONOLOBUS (WINGED BEAN)

BEANS

TUBERS

IPOMOEA AQUATICA (KANGKONG)

ROSA RUGOSA (ROSE HIPS)

IPOMOEA BATATAS (SWEET POTATO)

GOURDS

to stabilise slopes. Mostly grown in the USA for hay, cut before flowers bloom. L. st*ipulacea* and *L. striata* are annuals.

LEUCAENA (*Leucaena leucocephela*)
A fast-growing, leguminous tropical tree to 10-20m (although can be kept to a manageable size if coppiced or grazed by cattle). Does best on well-drained soils. Contains a mimosine that may cause toxicity in stock if overfed; a low mimosine variety is *L. leucocephela* var. *Cunninghamii*. Also, CSIRO scientists have isolated a microbial culture which cattle can use to break down the toxic substance in their stomaches. As long a leucaena is kept to 30%-40% of diet, there are no adverse effects even from normal leucaena varieties.

USES: Excellent high-quality forage (both leaves and pods) for cattle, sheep, goats; palatable and nutritious. Can be cut-and-fed, or stock let in to browse. Also useful in revegetating tropical hillslopes prone to erosion. Excellent coppicing for firewood, good timber. Rich in organic fertiliser; used as mulch in alley cropping. Fixes nitrogen in the soil. Used extensively as living fence/hedge species in West Africa and India.

LOQUAT (*Eriobotrya japonica*)
A small evergreen tree to 7m. Slow to develop from seed; use proven cultivars grafted onto loquat, pear, or quince stock. Yields by 6th year, peaks in 15-20 years. Suited to temperate areas; needs sheltered, sunny position. Frost-hardy but needs warmth for fruiting. Suits most soils, but is a gross feeder (plant near leachline outfall). USES: Fresh fruit in spring; medium understorey tree. Poultry and pig fodder (fruits).

LUCERNE/ALFALFA (*Medicago sativa*)
Perennial, leguminous herb with life expectancy of 10 years. USES: Human food: foliage as alfalfa tea; alfalfa sprouts for salads. Major temperate animal fodder plant. Excellent bee forage, blooming just after sweet clover. Soil improver, drawing up subsoil nutrients; useful ground cover/living mulch under trees.
Also **TREE MEDIC** (*Medicago arborea*): Perennial leguminous shrub to 4m; grows in temperate zones. Important fodder shrub with foliage equivalent to lucerne. Can be netted and sheep allowed to browse.

MACADAMIA (*Macadamia tetraphylla, M. integrifolia*)
Slow-growing, evergreen nut tree to 20 m; subtropical to tropical climates. Need windbreak protection. Grafted varieties bear in 6-7 years. Native to Australia, grown extensively in Hawaii and California.

USES: High-value nuts, difficult to crack by hand. Nut shells make excellent mulch. As with many trees, can be grown in pasture, with sheep let in after trees are mature enough to withstand grazing animals.

MAPLE (*Acer saccharum, A. macrophyllum*)
Deciduous cold area tree to 30m. Long-lived to over 200 years. Tolerates partial shade. Sends out growth inhibitor to nearby plants through roots. USES: Maple sugar, tapped in winter. Ornamental: red and yellow autumn leaves. Good carving wood. Bee forage.

MESQUITES (*Prosopis juliflora, P. tamarugo*)
Leguminous spreading shrubs and small trees 10-15m. Arid climates; totally drought-resistant and extremely salt tolerant. Grown from saline desert to semi-desert zones. *P. juliflora* (honey mesquite) yields 50 tons of pods per hectare, with 3-5 years to production. Caution: Easily becomes rampant.

USES: Major fodder trees of drylands for stock and poultry; 14cm-long pods are high in sugar, some protein. Pods made into a syrup (in Peru). Bee forage. Coppices easily for firewood. Also *P. alba, P. nigra, P. pallida* and *P. chilensis*.

MORINGA (*Moringa oleifera*)
Also called the horseradish or drumstick tree. Small tree to 10m; propagated by cuttings. Tropical, fast-growing. Tender pods as vegetables; flowers and young leaves eaten. Fried seeds. Roots as condiment (like horseradish). Twigs and leaves lopped for stock fodder.

MULBERRY (*Morus* spp.)
Deciduous dome-shaped trees to 20m, grow from temperate to subtropical climates. Main species are black mulberry (*M. nigra*), red mulberry (*M. rubra*) and white mulberry (*M. alba*). Can be grown in full sun but is also shade tolerant. Easily grown from seed or cuttings.

USES: Edible berries, *M. nigra* and *M. rubra* have superior fruit. *M. alba* is fast-growing, with short fruiting season; leaves are used as silk-worm food in China. Excellent trees for poultry and pig forage as fruits are numerous and fall easily to the ground. Leaves can also be fed to cattle. Useful wood for fenceposts and barrels.

NASTURTIUM (*Tropaeolum majus*)
A creeping or climbing perennial, usually grown as an annual; frost-sensitive. Prolific in moist gardens, but will also grow in most soils and sites. USES: Good ground cover and companion plant around fruit trees. Seeds can be pickled as a substitute for capers; they are also used medicinally as an antiseptic. Leaves and flowers edible in salads.

NATAL PLUM (*Carissa grandiflora*)
Thorny, evergreen shrub to 2m; grows in dry subtropics/tropics. Ripe fruits eaten raw; preferably made to conserves. Substitute for cranberry sauce. Attractive ornamental shrub; valued as a hedge in South Africa.

HEDERA HELIX (ENGLISH IVY)

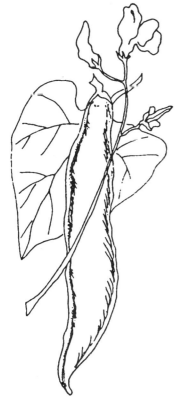

PHASEOLUS COCCINEUS
(SCARLET RUNNER BEAN)

OAKS (*Quercus* spp.)

Mostly large, spreading, deciduous trees up to 40m, although some are smaller or even prostrate. Long-lived; many are fast growing and bear acorns early. Large habitat range from dryland soils to acid swamps; temperate to subtropical climates (most species are well-suited to cold areas). Good germination, although acorns sometimes lose viability in a year. Yield is variable, usually yielding in alternate years.

USES: Acorns as animal forage, high carbohydrate. Most valuable for pigs, although crushed acorns and leaf mould are fed to poultry. Species used are "sweet", or low in tannin. Excellent hardwood timber and firewood. Some species used for wine vats to aid maturation process. Oaks offer shelter for stock and are good fire sector species (poor burners when "green"). Leaves are used for animal bedding. Following are a list of some species suited to particular uses:

Human food: Acorns contain tannin which can be removed from ground acorn meal by leaching in streams and cooking. Some sweeter acorns are: *Q. ilex* var. *ballota* (a cultivar of the holm oak) which is the best old world eating acorn used in Portugal and Spain. *Q. alba* (white oak) a common North American tree with acorns boiled like chestnuts by Native Americans.

Fodder: Best is *Q. ilex* (holm oak) and *Q. suber* (cork oak); mixed stands are grown in Portugal for pig forage, with very high yields on alternate years. Such mixed oak forests yield 68kg/ha per year over a ten year period. Other fodder species are *Q. prinoides* (chinquapin oak), *Q. alba* (white oak), and *Q. minor* (post oak).

Timber: Most oak trees produce superior quality timber. Some important species are *Q. robur* (English oak) used for centuries in buildings and ships; *Q. petraea* (durmast oak); *Q. alba* (white oak), also used for barrel making; and *Q. rubra* (red oak), used extensively for furniture.

Cork: *Quercus suber*, the cork oak, is cultivated in Spain and Portugal for wine/champagne bottle stoppers, insulation, flooring, etc. Once mature, cork can be harvested every 8-10 years without harming the tree. A hectare of cork oak will yield an average of 240 kg/year.

Other Uses: *Q. mongolica* is the host plant for the tusser silkworm of China and Japan; these are semi-domesticated and produce a high-quality silk. *Quercus velutina* (black or quercitrin oak) yields a permanent yellow dye from its inner bark. *Q. ilex* and *Q. alba* are used for high-quality charcoal production.

OAKS

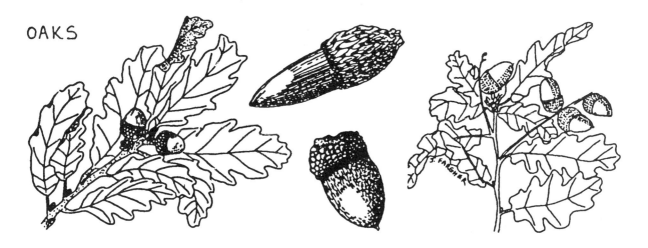

OLIVE (*Olea europaea*)

Small, evergreen tree to 8m; long-lived (up to 700 years). Dryland plant of Mediterranean region, not suited to maritime or cold regions (although fairly frost-hardy, fruit needs hot summers to ripen). Propagation by cuttings; olives bear in 4-6 years. Can grow on thin, rocky soils but yield is best on fertile soils.

USES: Fruit is eaten green or ripe; green olives must be soaked in a lye solution before pickling to remove bitterness. Excellent oil crop: fruit picked when fully ripe (but not soft), then crushed to a mash and placed in cloth bags. These are pressed and the oil collected. Good olive varieties yield as much as 30% oil. The remaining pulp after pressing can be fed to stock. Olive trees are a good shelter and occasional forage for stock.

PALMS

Woody perennials with many uses, from human food, oils, sugar, animal fodder, structural material, thatch, and fibre. Most useful palms grow in dry or wet tropics. Have deep tap roots, and many are successfully used in agroforestry (crops and pasture) as they do not compete for water.

Date palm (*Phoenix dactylifera, P. sylvestris, P. canariensis*): Dioecious, need one male to 60-80 females. Staple food yielding dates; old trees are tapped for toddy (sugar). Inferior species of dates can be used for animal fodder or possible fuel crop.

Borassus palms (*Borassus* spp.): Palmyra (*B. flabellifer*) tapped for sugar in India (produces 170 pounds of nectar per acre, or 40,000 litres of alcohol fuel). Timber is hard and durable. Others are *B. aethiopican, B. sundaicus*.

Doum or **gingerbread palm** (*Hyphaene thebaicus*): Multi-stemmed, branched palm to 15m, bearing heavy crops of edible hard-shelled fruits. Staple food and fodder crop of arid lands, mainly Egypt.

Coconut (*Cocos nucifera*): Essential plant of many tropical island cultures. Yields coir for rope, thatch, oil, drinking "water", nut meats, and sugar from flower stalks.

Chilean wine palm (*Jubaea spectabilis*): Temperate-zone palm yielding 410 litres of sweet sap annually. Cold hardy. Fruits with edible nut, useful for fodder.

Peach palm or **Pejibaye** (*Bactris (guilielma) gasipaes*): A spiny-trunked plant; staple plant of Central & South America exceeding maize in protein and carbohydrate yields per acre. Fruits chestnut-like, boiled and dried as food. Also for poultry and pig forage. Hardy only in frost-free areas.

PASSIONFRIUTS (*Passiflora* spp.)

Evergreen perennials; vigorous growers (sometimes rampant as they will naturalise and climb in forest trees). USES: Edible fruit, poultry and pig fodder, sun deflector to shade walls, used to cover (and keep cool) water tanks and sheds. Ornamental, with showy flowers.

Black passionfruit (*Passiflora edulis*) is a vigorous climber of subtropical to tropical areas. Cultivated on fixed fence trellis, cropping for 4-8 years (some varieties last longer). Frost tender in early growth.

Banana passionfruit (*P. mollisima*) is grown in temperate maritime climates; will withstand moderate frost once established. Yields from late autumn to early summer, and is a valuable poultry fodder (fruit seeds). An under-used fruit for winter fresh fruit, more easily peeled than *P. edulis*.

Lillikoi (*P. alata*) is a hardy, vigorous grower of the subtropics and tropics; plant two or more for best cross-fertilisation. Delicious fruit.

Other edible passionfruits of the tropics are granadilla (*P. quadrangularis*), sweet granadilla (*P. ligularis*), and waterlemon (*P. laurifolia*).

PAULOWNIA (*Paulownia tomentosa, P. fargesii*)

Quick-growing, drought-resistant deciduous trees to 15m. Mild temperate to subtropical range, with *P. fargesii* in the cooler climates. Grown extensively in China. Has deep taproot and will not compete with pasture, crops. Has large leaves, but with some pruning and wide spacing allows light through.

USES: Timber crop for fine furniture, boxes, chests. Used in agroforestry to shelter cereal, soybean, and cotton

SUB DOMINANT: COFFEE, CACAO, VANILLA, PIGEON PEA ....
DOMINANT SPECIES: AVOCADO, COCONUT, JAKFRUIT, CASHEW, PECAN .... Coconut Circle

crops; wood taken in 6-12 years (pruning and shaping necessary to maintain good log growth). Leaves contain nutrients, nitrogen; can be used as stock fodder and mulch.

PERSIMMON (*Diospryros kaki, D. virginiana*)

Many varieties, especially in Japan. Deciduous tree to 15m, yielding fruit in winter. Temperate to subtropical climates. Fairly frost-hardy; does well in most well-drained soils. Japanese persimmon (*D. kaki*) does best in full sun, while American persimmon (*D. virginiana*) can tolerate partial shade.

USES: Fruit, eaten when over-ripe (harvested when hard and ripened indoors). Fallen fruit is an excellent pig and stock food. Ornamental plant, with autumn colour (spectacular red fruits on leafless tree). A good front yard plant, along with other such ornamental edibles as nasturtium, kale, almond, peach, currant, etc.

PIGEON PEA (*Cajanus cajan*)

Leguminous woody shrub of dry subtropics and tropics; frost sensitive. Quick-growing, short-lived perennial; sometimes grown as an annual. 1-4m tall.

USES: Major tropical food grain, green seeds and pods used as vegetables. Ripe seeds for flour, dhal, sprouts (22% protein, 10% calcium). Important forage plant eaten green or made into hay or silage. Sometimes planted in pastures as a browse plant. Ideal windbreak and shade for vegetables; leaves cut for mulch on garden beds. Shade tree plantations (coffee, cacao) and vanilla production in India. Useful windbreak hedge species.

Used in Asian medicine as a treatment of skin irritations, cuts. Leaves used for silkworm culture in Malagasy. Green manure and cover crop. Used in erosion control. Dried stalks for firewood, thatching and baskets in India.

PRICKLY PEAR (*Opuntia* spp.)

Spiny cacti with flat, fleshy pads grown in dry subtropics/tropics. Like full sun; grow to 2m. Propagated by planting pads into the ground. Will grow in poor soils; drought-resistant. **Caution:** can be invasive; birds carry seeds.

USES: Fruit, eaten fresh or stewed (numerous hard seeds); use gloves to harvest, then scrub off fine spines and peel. Seeds are nutritious and are sometimes ground for animal feed. Young *Opuntia* pads are de-spined and sold in Mexican, Indian markets for human food; pads also fed to stock (spines are burnt off). Good barrier hedges. Some varieties are: mission prickly pear (*O. mega-cantha*); common prickly pear (*O. ficus-indica*); *O. undulata, O. streptacantha*.

PRUNUS SPP.

These deciduous species contains some of the most important temperate fruits: apricot, plum, almond, peach, nectarine, cherry. Many cultivars, some miniature varieties. Most are small trees and shrubs 1-10m tall. Mediterranean climates, warm dry summers best. Semi-tolerant of drought.

USES: Mainly for fruit, usually eaten fresh or in conserves, juice. Almonds are a storable product. Some species such as damson plum (*P. instilia*), sour cherry (*P. cerasus*), and common plum (*P. domestica*) will form thickets, making an excellent hedge for windbreak, wildlife habitat. All species good bee forage.

QUEENSLAND ARROWROOT (*Canna edulis*)

A clump-forming perennial of the subtropics and tropics (originally from the Americas). One of the hardiest arrowroot plants, can grow in temperate areas where there is little frost (needs warm, sunny position).

USES: Tubers cooked for a sweetish taste, though inferior to sweet potato due to fibre. Arrowroot flour. Animal forage, especially pigs. Also used as a garden windbreak and weed barrier with comfrey and lemongrass; and can be chopped occasionally for garden mulch.

QUINOA (*Chenopodium quinoa*)

Hardy annual to 1-2m, grown in South American Andes; cold temperate, dryland. Drought-tolerant. Sow in spring

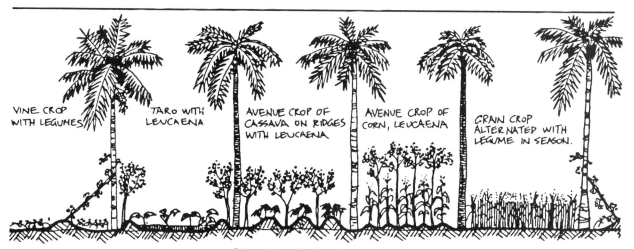

Classical Palm Intercrop

after frost. Nutritious grain food, tasty greens. Grind seeds into flour or remove bitterness of whole grains by soaking; use boiled or in soup. Chicken fodder.

Other useful *Chenopodium* spp. for human food and chicken fodder are fat hen or lamb's quarters (*C. album*) with calcium-rich leaves for salads, seeds relished by poultry and birds; and good king henry (*C. bonus-henricus*), young plant eaten like asparagus and spinach.

ROSELLA (*Hibiscus sabdariffa*)
Fast-growing annual shrub of subtropics and tropics. Grows 1.5-2m tall. Tolerates most soils; must be well-drained. Needs long summer growing period.

USES: Fruits stewed or used in desserts and drinks; conserves. Tender leaves and young stems used as a salad or steamed; leaves are chopped as a savoury herb (for curries). Another useful plant of the *Hibiscus* family is **okra** (*H. esculentus*) with the tender pods boiled or sliced and fried. Used in soups and gumbos.

SALSIFY (*Tragopogon porrifolius*)
Temperate, biennial clumping plant to 0.6m, often planted as an annual. Cultivated for its oyster-flavoured edible tap-root (harvested autumn, winter, spring). Young leaves and flowers edible in spring and summer.

SCARLET RUNNER BEAN (*Phaseolus coccineus, P. multiflorus*)
Herbaceous perennial (grown as an annual in cold climates) with thick root stock. Tolerates some frost; grown in mild coastal or island climates. Needs cool periods to fruit heavily.

USES: Edible young pods, beans fresh or dried. Good trellis plant for shade; bright red ornamental flowers. Tubers are boiled as a vegetable in Central American highlands. Other useful *Phaseolus* are tepary bean (*P. acutifolius*), a high-value dryland species; and lima bean (*P. lunulatus*), a tropical low hedge plant on fences.

SESBANIA (*Sesbania bispinosa, S. aculeata, S. grandiflora*)
Fast-growing (4-6 m/year), short-lived subtropical and tropical legume tree 6-9m high. Drought-resistant. Easily propagated by seed.

USES: Seeds used for poultry fodder and leaves for forage. *S. aculeata* used in Asia as traditional green manure crop and border plant (nitrogen-fixer) planted together with rice. *S. grandiflora* grown in the Mekong delta in home gardens for its leaves and flowers used for human food and livestock & poultry. Planted along rice paddies, yield up to 55 tons of green material per hectare. Used as temporary shade trees in nurseries. Windbreak in citrus and coffee, banana. Living fence and firewood source. Used for large-scale reafforestation of bare land outside forests in Indonesia.

SIBERIAN PEA SHRUB (*Caragana* spp.)
Tall, leguminous shrubs 1-5m, forms thickets. *Caragana arborescens* is the only species that grows into a tree. Very cold and wind hardy, growing from arctic circle to warm, dry climates. Seeds burst out of 6cm-long pods and should be collected in bags before completely ripe if needed for seed.

USES: Windbreak and hedge shrub for very cold climates. Seeds are excellent poultry forage food, and pods can be left on the shrubs to burst open. Wildlife habitat, sheltering small animals in the thickets. *C. arborescens* leaves produce a blue dye. Nitrogen-fixer.

STONE PINE (*Pinus pinea* & other spp.)
Conifer up to 10-30m tall; slow-growing and long-lived. Suits cool areas and can grow on exposed, dry, rocky sites.

USES: Pine nuts or kernels are rich in oil, have a very good flavour. Cones are collected when mature but unopened; opened in summer sun or dryer and nuts are shaken free. Many species have excellent edible nuts, including pinyon pine (*P. edulis*), Coulters pine (*P. coulteri*), *P. cembra* (Europe), *P. gerardiana* (Afghanistan).

SUNFLOWER (*Helianthus annus*)
Annual plants 0.7-3.5m tall; temperate to tropical climates (not suited to the wet tropics, however). Drought-resistant, but do best when watered at intervals. Grow on a wide range of good draining soils. Release root exudate; some crops do not grow next to them.

USES: High-value protein seed for human and livestock, especially poultry, pigeons. Whole heads may be given to stock. Salad and cooking oil made from seeds; with seedcake residue fed to stock. Also used in blends with linseed for paints and varnishes. Lubricant and lighting. Stalks and hulls are mulch, bedding for livestock.

SUNN HEMP (*Crotalaria juncea*)
Tall shrubby annual 1-3m subtropics and tropics; frost-sensitive. Quick-growing, large leaved legume. Hardy and drought-resistant.

USES: Cultivated for fibres used as twine, paper, nets, sacking (better than jute). Root exudate said to control nematodes in the soil. Easily grown in gardens, with leaves used for mulch. *Crotalaria brevidens* used as annual fodder in tropical Africa. Green manure crop, often grown in rotation with rice, maize, cotton; and interplanted with coffee, pineapple. When thickly-sown, will smother all weeds, even vigorous grass weeds.

SUNROOT (*Helianthus tuberosus*)
Also called Jerusalem artichoke. Tall perennial which dies back to roots; 1-3m tall. Propagated by tubers. Yields are often 4-5 times that of potatoes. Hardy, wide climatic range from temperate regions to tropics. Will tolerate

poor soils, drought. Like sunflowers, sunroots release a root exudate which is toxic to some plants.

USES: Human food; tubers eaten as vegetable. Animal forage: dry stalks and leaves eaten by goats; tubers by pigs. Fast garden windbreak; also useful to break up hard soils. Leaves used for mulch in gardens after tubers harvested.

SWEET POTATO (*Ipomoea batatas*)
Perennial twining plant, often treated as an annual. Temperate to tropical tubers usually planted on ridges or mounds (cannot stand waterlogging). Propagated by stem cuttings in tropics; tuber sprouts in temperate climates. Needs frost-free growing period of 4-6 months.

USES: Important food source, eaten boiled or baked. Used for canning, drying, flour manufacture, and as a source of starch, glucose, syrup and alcohol. Also fed to livestock. Vines are widely used as fodder for stock. Grown in subtropics as a groundcover for orchards, but must occasionally be slashed from tree trunks. Die back in frost.

TARO (*Colocasia esculentus*)
Tropical wet culture plants with over 1000 cultivars. Grown either in wetland terraces with *Azolla* fern (for nitrogen fixation) or on mulched and irrigated plots. Staple food of the tropics. Large root is eaten, although some taros are grown for their leaves. The leaves of many taros are poisonous.

TAUPATA (*Coprosma repens*)
Also called New Zealand mirror plant. Large, evergreen shrub 2-3m with shiny leaves; dioecious (separate male and female plants). Easily grown from cuttings. Temperate climates; windhardy and resistant to salt spray, drought, and fire. Common ornamental seaside plant in New Zealand and Tasmania.

USES: Hedgerow plant and fire retardant. Fruit and seeds are excellent poultry forage. Leaves are eagerly eaten by sheep, horses, cows. Pruned clippings make a good mulch or compost.

TAGASASTE (*Chamaecytisus palmensis*)
NOTE: Previously named tree lucerne (*Cytisus proliferus*). Nitrogen-fixing legume tree 6-10m, living for more than 30 years. Easy to grow from seed (scarify or pour boiling water over seeds and soak). Tolerant of poor soils, drought, wind; originated in dry, Mediterranean-type climate, but does well in cool temperate areas, withstanding light frosts. Tagasaste recovers after pruning or defoliation by animals.

For best results fertilise with trace elements and lop branches regularly (either by hand or browsing) to give a more bushy foliage. Seed can be direct drilled into pasture, but plants should be protected from stock for up to 3 years (or stock let in for brief periods to graze). If sheep ringbark trees, cut to the ground to encourage new growth; this will form thickets more resistant to sheep damage.

USES: Foliage an important protein-rich fodder for stock during drought and at the end of summer. Bee forage; many small white flowers. Chickens eat seeds. Windbreak hedge. Nurse plant surrounding frost-sensitive trees in early years. Excellent cut mulch; tree can be lopped 3-4 times in summer.

TAMARILLO or TREE TOMATO (*Cyphomandra betacea*)
Short-lived shrub to 3-6m, of the tomato family. Sown from seed or propagated by cuttings from 1 or 2 year old wood. Yields in two years. Subtropical, marginally suited to cool areas (place in a sheltered, sunny position—will tolerate mild frost). Well-drained soil. USES: Fruit high in vitamin C; used fresh, stewed, conserves. Commercially grown in New Zealand; high-value crop.

TRAPA NUT (*Trapa natans, T. incisa*)
Also called Indian water chestnut. Several species, temperate to tropical regions. Aquatic perennial, floats in water 2-3 feet deep. Needs high nutrients. USES: Important starchy food plant, rich in iron; flour like arrowroot.

WALNUT (*Juglans regia, J. nigra*)
Spreading, deciduous trees to 30m; long-lived. Temperate climate, cold areas. Yields best on deep, well-drained rich soils. Release root exudate which inhibits some understorey plants, although pasture does well. USES: Both species are important for nut production, timber, specialty woods. Husks produce a dye. Black walnut (*J. nigra*) rootstock is resistant to *Armillaria* root rot; all commercial English walnut stands are grafted. Black walnut is a particularly sought-after wood, with very high prices paid for good, straight timber (yields in 40-50 years).

WHITE CEDAR (*Melia azedarach*)
Short-lived (20 years), deciduous tree 9-12m tall. Suited to a wide range of warm climates (tropics to Mediterranean climates, e.g. South and West Australia).

USES: Fast-growing shade tree; good for afforestation. Valuable timber; resistant to termite attack (does not need to be treated) and used for poles, furniture, and roofing material. Fuelwood. Coppices well; trees lopped for green manure. Leaves, bark, and fruits are credited with insect-repellent qualities. Extracts of the leaves are used as a spray against grasshoppers, and leaves placed in books and wool clothing to protect against moths.
Caution: fruits are very poisonous.

WILLOWS (*Salix* spp.)
Around 300 species. Mainly spreading, deciduous trees;

water-loving. Mostly temperate climate. Easily propagated from stem cuttings. May become naturalised or rampant, especially along streams.

USES: *Salix viminalis* (osier willow) and other species used for basketry. Long 1-2 year old shoots are cut from pollarded willow stumps, or from thickets of willow stems (trunk cut at ground level). *S. matsudana* is used in New Zealand for erosion control. Weeping and pussy willows (*S. discolor*), among others, are excellent bee forages. Willows are fire retardants (steam rather than burn). *S. matsudana* var. *Tortuosa* has lush foliage for emergency sheep and deer fodder during drought; one hectare of willows can maintain 1000 sheep for 6 days (data from *Agroforestry in Australia and New Zealand*).

WINGED BEAN (*Psophocarpus tetragonolobus*)
Leguminous, twining vine, growing to over 3m when supported. Valuable, nutritious tropical garden bean.

USES: Edible pods, young leaves, shoots, flowers as vegetables; immature tuberous roots eaten raw or cooked. Very high protein content. Can be used as for soy beans for processing to bean cake. Seeds contain oil for cooking, soap, and lighting. Dry flowers eaten like mushrooms. Excellent nitrogen-fixer (heavy nodulation), soil conditioner and cover crop for the tropics.

YAM BEANS (*Pachyrrhizus erosus, P. tuberosus*)
Herbaceous, twining plant 2-6m tall. Warm-climate, dryland perennial beans with crisp, edible tubers; harvested after 4-8 months. Mature seeds and leaves toxic.

USES: Tubers widely eaten in Mexico, Philippines, SE Asia, raw or cooked. Called jicama (*P. erosus*) in Mexico and eaten in salads or sliced thinly and sprinkled with salt, lemon juice and chilli sauce. Young pods of *P. erosus* sometimes eaten like French beans. Old starchy tubers are fed to cattle.

YARROW (*Achillea millefolium*)
Herbaceous perennial to 1 m, with white flower heads. Drought-resistant; naturalises along roadsides and disturbed soils. Bee forage. Insectary plant (a member of the composite family which attracts beneficial insects). Flowering tops and foliage of medicinal use for stock, especially sheep.

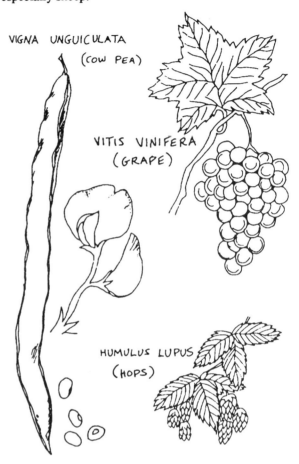

VIGNA UNGUICULATA (COW PEA)

VITIS VINIFERA (GRAPE)

HUMULUS LUPUS (HOPS)

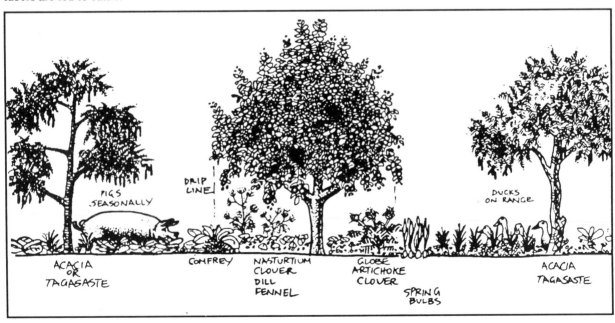

ACACIA OR TAGASASTE — PIGS SEASONALLY — COMFREY — DRIP LINE — NASTURTIUM CLOVER DILL FENNEL — GLOBE ARTICHOKE CLOVER — SPRING BULBS — DUCKS ON RANGE — ACACIA TAGASASTE

Appendix B

SPECIES LISTS IN USEFUL CATEGORIES

The following are lists of useful permaculture categories, with no attempt to describe particular plants. Some of those listed can be found in Appendix; others are so common as to need no description. Asterisk (*) marks tropical/subtropical species.

Table 1: Plants with Food Products from Roots, Tubers, or Shoots

*Arracacha	Asparagus
Bamboos	Beet
*Cassava	Carrot
Celeriac	Chicory
Choko	Dandelion
*Yam beans	Onion
Parsnip	Radish
Sunroot	Potato
*Taro	Turnip
Salsify	*Queensland arrowroot
Peanut	Scarlet runner bean
Duck potato (USA)	

Table 2: Plants Giving Storable Food Products

A. Nuts

Almond	Black walnut
Walnut	*Bunya pine
Butternut	Chestnut
Filbert, hazel	*Macadamia
Ginkgo	Pecan
*Pistachio	Oaks
Stone pine & other pine nuts	

B. Fruits (suitable for local drying & storing)

Apple	Apricot
Fig	Jujube
Peach	Prune plum
Cherry	Pear
*Mango	*Pineapple
*Banana (some small varieties)	
Grape (some raisin varieties)	

C. Flours and meals

Carob	Honey locust
Sweet chestnut	White mulberry
Pigeon pea	
*Indian water chestnut	
*Queensland arrowroot	

D. Cooking and salad oils

Almond	Beech	
Hazel	Olive	
Live oak	Walnut	
Mustard	Grapeseed	
Safflower	Rape	Sunflower

Table 3: Fresh Fruits

A. Temperate

Alpine strawberry	Loquat
Apple	Medlar
Apricot	Mulberry
Blueberry	Nectarine
Cape gooseberry	Peach
Checker berry	Pear
Kiwifruit	Persimmon
Feijoa	Plum
Fig	Grape
Strawberry	Grapefruit
Strawberry guava	Jujube
Cherry	Tamarillo
Berries (black, logan, boysen, red)	
Banana passionfruit	Black, red currants

B. Subtropical/tropical

Mango	Jakfruit
Guava	Rambutan
Carambola	Mangosteen
Lychee	Naranjilla
Sapote	Jaboticaba
Mammey sapote	Pepino
Papaya	Custard apple
Prickly pear	Natal plum
Granadilla	Pineapple
Passionfruit	Citrus spp.

Table 4. Fruit Used in Cooking, Preserves, Wine

Cranberry	Elderberry
Quince	Huckleberry
Cornelian cherry	Barberry
Pomegranate	Cumquat

Table 5. Fruit High in Vitamin C

*Barbados cherry	Citrus
Rose (*Rosa rugosa*)	*Rosella
*Guava	

Table 6. Animal Forages and Feeds

A. Nuts, pods, seeds

Almond	Mesquites
Beech	Taupata
Oaks	Siberian pea shrub
Honey locust	Tagasaste
Hazel	Walnut

Hickories	Acacias
*Ice cream bean	*Leucaena
*Pigeon pea	Amaranth
Quinoa	*Sesbania
*Winged bean	Carob

B. Foliage

Bamboo	Lespedeza
Chicory	Lucerne
Comfrey	Lupin
Tagasaste	Pampas grass
Sunroot	*Vigna* spp.
*Lab-lab bean	*Leucaena
Tree medic	*Pigeon pea
*Sesbania	Taupata
Willow	Kurrajong
*Winged bean	Dandelion
Choko/chayote	

D. Roots, tubers, rhizomes

* Arracacha	*Yam beans
Sunroot	Chickory
*Yam	Choko
Comfrey	Sweet potato
Arrowhead, duck potato	
*Queensland arrowroot	

Table 7. Edible Flowers For Salads

Daylily	Borage
Calendula	Feijoa
Black locust	Nasturtium
*Sesbania	Dandelion
Salsify	Sweet violet
Zucchini	*Winged bean
Rose (*Rosa rugosa*, *R. canina*)	

Table 8. Hedge Plants

Taupata	Some clumping bamboos
Alder	Hawthorn
Hazel	Russian olive
Autumn olive	Elderberry
Laurelberry	Pampas grass
Coprosma	*Queensland arrowroot
Pomegranate (closely-spaced; clipped)	
Some *Prunus* spp. (Damson plum, sour cherry)	

Table 9. Animal Barrier Plants

(Spiny or unpalatable dense thickets)

Euphorbia spp.	Hawthorn
Gorse	Sloe
Honey locust	Natal plum
Prickly pear cactus & other cactus spp	

Table 10. Useful Perennial Vines

A. Deciduous

Grape	Wisteria
Kiwifruit	Scarlet runner bean
Scarlet trumpet vine	*Yam beans
Virginia creeper	

B. Evergreen

*Passionfruit	Choko/chayote
*Vanilla	*Lab-lab bean
Jasmine	Ivy

Table 11. Pest Control Plants

*Sunn hemp (nematodes)
Marigold (*Tagetes* spp.) nematodes
Pyrethrum daisy (broad spectrum insecticide)
White cedar and neem tree (insecticide)
Tobacco (insecticide)
Derris root (*Derris elliptica*)
Rhubarb (insecticide)

Table 12. Umbelliferous Plants

Celery	Angelica
Florence fennel	Parsley
Dill	Chervil
Lovage	Queen Anne's lace
Caraway	Coriander
Fennel	Cumin
Anise	Sweet cicely
Parsnip	Carrot

Table 13. Composite Plants

Tarragon	Southernwood
Tansy	Chamomile
Wormwood	Daisies
Artichoke	Salsify
Sunroot	Sunflower

Table 14. Water or Wetland Plants

Azolla	Rush (*Scirpus* spp.)
Watercress	Water chestnut
Mint	*Kang kong
Water lily	*Lotus
Wild rice	Rice
Duckweed	Duck potato (arrowhead)
Willows	Cranberry
Highbush cranberry	
Cumbungi or cattail	
Reed (*Phragmites* spp.)	

Table 15. Bee Forage

Almond	Lavender
Apple	Loganberry
Bergamot	Lucerne/alfalfa

Blackberry	Lupin
Black currant	Mesquites
Apricot	Mints
Black locust	Borage
Leatherwood	Cherry plum
Clover	Comfrey
Peach	Dandelion
Pear	Tagasaste
Raspberry	Gooseberry
Rosemary	Citrus spp.
Sage	Hawthorn
Sloe	Hyssop
Sour cherry	Laurelberry
Pride of Madeira	Some Eucalypts
Osier willow (& other willows)	

Table 16. Species for Very Dry Sites

Almond	Mesquites
Black locust	Mulberry
Burr oak	Olive
Carob	New Zealand spinach
Cork oak	Pampas grass
Tagasaste	Prickly pear
Many *Acacia* spp.	*Pistachio
Pomegranate	Jujube
Fig	Quandong
Holm oak	Rosemary
Honey locust	Stone pine
Lavender	Taupata
Most aromatic herbs	

Table 17. Legumes & Other Nitrogen-fixing Plants

The trees can be coppiced for green manuring and animal fodder. * denotes non-legume nitrogen-fixer.

A. Temperate

Trees

Tagasaste	Black locust
Autumn olive	Russian olive
*Alder	Siberian pea shrub
Albizia	*Ceanothus
Tree medic	

Small species

Azolla (aquatic)	Fenugreek
Clover	Lucerne
Beans & peas	Vetch
Lupin	Lespedeza

B. Warm/dry climates

Trees

Mesquite	Acacias
Tagasaste	Albizia
Casuarina	

C. Tropics/Subtropics

Trees

Acacias	Albizia
Gliricidia	Calliandra
Leucaena	Sesbania
Pongamia	Tamarind
Cassia	Ice cream bean tree
Tipuana tipu	

Small species

Pigeon pea	Lab lab bean
Winged bean	Peanut
Beans & peas	Clover
Lucerne	

REFERENCES

BOSTID, *Tropical Legumes: Resources for the Future*, National Academy of Sciences, Washington DC, 1979.

Brouk, B., *Plants Consumed by Man*, Academic Press, NY, 1975.

Douglas, J. Sholto, *Alternative Foods*, Pelham Books Ltd., 1978.

Hedrick, U.P. (ed), *Sturtevants' Edible Plants of the World*, Dover, NY, 1972.

Masefield, et alia., *The Oxford Book of Food Plants*, Oxford University Press, London, 1969.

Mollison, Bill and David Holmgren, *Permaculture One*, 1978, Tagari Publications.

Lindegger, Max O., Subtropical Fruits - A Compendium of Needs & Uses, 1984, Permaculture Consultancy, 56 Isabella Ave., Nambour QLD 4560.

Litwin, Shery, Plant Species Index in *The Future is Abundant: A Guide to Sustainable Agriculture*, Tilth, 1982.

Usher, George, *A Dictionary of Plants Used by Man*, 1974, Oxley Printing Group, United Kingdom.

Appendix C

COMMON AND LATIN PLANT NAMES

A: PLANTS MENTIONED IN TEXT BY COMMON NAME

Acacia, *Acacia spp.*
  Blackwood, *A. melanoxylon*
  Brisbane wattle, *A. fimbriata*
  Cootamundra wattle, *A. baileyana*
  Golden wattle, *A. longifolia, A sophorae*
  Green wattle, *A. mearnsii*
  Kangaroo thorn, *A. armata*
  Mulga, *A. aneura*
  Raspberry jam acacia, *A. acuminata*
  Tropical acacias, *A. auriculiformis* + spp.
  Silver wattle, *A. dealbata*
  Weeping wattle, *A. saligna*
  White acacia, *A. albida*
African marigold, *Tagetes erecta, T. minuta*
Agave, *Agave spp.*
Albizia, *Albizia spp.*
Alder, *Alnus spp.*
Alfalfa, *Medicago sativa*
Aloe, *Aloe spp.*
Alyssum, *Alyssum spp.*
Amaranth, *Amaranthus spp.*
Anise, *Pimpinella anisum*
Angelica, *Angelica archangelica*
Apple, *Malus pumila*
Apricot, *Armeniaca vulgaris*
Arracacha, *Arracacha esculenta*
Arrowhead, *Sagittaria spp.*
Arrowroot (Queensland), *Canna edulis*
Arrowroot (West Indian), *Maranta arundinaceae*
Asparagus, *Asparagus officinalis*
Asparagus fern, *Asparagus setaceus*
Autumn olive, *Elaeagnus umbellata*
Avocado, *Persea americana*
Azolla, *Azolla spp., A filicoides*
Bamboo, *Bambusa, Phyllostachys, Arundinaria, Dendrocalamus,* and allied genera
Banana, *Musa paradisiaca* + spp.
Banana passionfruit, *Passiflora mollissima*
Banksia, *Banksia spp.*

Barley, *Hordeum vulgare*
Basil, *Ocimum basilicum*
Bayberry, *Myrica californica*
Bean, broad, *Vicia faba*
  common, *Phaseolus vulgaris*
  Dolichos, *Lab-lab purpureus*
  fava, *Vicia faba*
  Lab-lab, *Lab-lab purpureus*
  lima, *Phaseolus lunatus*
  mung, *Vigna radiata*
  soya, *Glycine max*
  yam, *Pachyrrhizos tuberosus*
  winged bean, *Psophocarpus tetragonolobus*
Beet, *Beta vulgaris*
Bell pepper, *Capsicum annuum*
Birch, *Betula spp.*
Bittermelon, *Momordica charantia*
Black locust, *Robinia pseudoacacia*
Black walnut, *Juglans nigra*
Blackwood, *Acacia melanoxylon*
Blueberry, *Vaccinium spp.*
Borage, *Borago officinalis*
Boysenberry, *Rubus ursinus*
Boxthorn, *Lycium ferrocissimum* + spp.
Brazilian cherry, *Eugenia brasiliensis*
Breadfruit, *Artocarpus altilis*
Brussel sprouts, *Beta oleracea var. gemnifera*
Buckwheat, *Fagopyrum esculentum*
Cabbage, *Brassica spp.*
Calendula, *Calendula officinalis*
Calliandra, *Calliandra spp.*
Carambola, *Averrhoa carambola*
Cape gooseberry, *Physalis peruviana*
Capeweed, *Arctotheca calendula*
Caraway, *Carum carvi*
Cardamon, *Elettaria cardamomum*
Cardoon, *Cynara cardunculus*
Cassia, *Cassia spp., C. multijuga*
Castor oil plant, *Ricinus communis*
Casuarina, *Casuarina spp.*
Cat's claw creeper, *Dexantha unguis-cati*
Catmint, *Nepetea cataria*
Cauliflower, *Brassica oloeracea*
Cedar, *Cedrus spp.*
Celery, *Apium graveolens*
Ceriman, *Monstera deliciosa*
Chamomile, *Chamaemelum nobile*
Chayote, *Sechium edule*

Cherry, *Prunus cerasus, P. avium*
Chervil, *Anthriscus cerefolium*
Chestnut, *Castanea spp.*
Chicory, *Cichorium intybus*
Chili pepper, *Solanum frutescens*
Chile jasmine, *Mandevilla laxa*
Chinese gooseberry, *Actinidia chinensis*
Chinese trumpet creeper, *Campsis grandiflora*
Chinese water chestnut, *Eleocharis dulcis*
Chinquapin, *Castanea pumila*
Chives, *Allium schoenoprasum*
Choko, *Sechium edule*
Citrus, *Citrus spp.*
Cleavers, *Galium aparine*
Climbing fig, *Ficus pumila*
Clover, *Trifolium spp.*
Cocoa, *Theobroma cacao*
Coconut, *Cocos nucifera*
Coffee, *Coffea spp. C. robusta, C. arabica*
Coriander, *Coriandrum sativum*
Cordia, *Cordia abyssinica*
Corn, *Zea mays*
Cotton, *Gossypium spp.*
Cowpea, *Vigna sinensis*
Cranberry, *Vaccinium marocarpon* + spp.
Crocus, *Crocus sativus*
Cross vine, *Bignonia capreolata*
Crotalaria, *Crotalaria spp.*
Cucumber, *Cucumis sativus*
Cumbungi, *Typha spp., T. angustifolia*
Cumin, *Cuminum cyminum*
Currants, Black, *Ribes nigrum*
  Gold, *Ribes aureum*
  Red, *Ribes rubrum*
Custard apple, *Annona spp.*
Cypress pine, *Callitris columellaris*
Daffodil, *Narcissus spp.*
Dahlia, *Dahlia spp.*
Daikon radish, *Raphanus sativus*
Dandelion, *Tarascum officinale*
Daylily, *Hemerocallis fulva*
Derris, *Derris spp., D. elliptica*
Dichondra, *Dichondra repens, D. micrantha*
Dill, *Anethum graveolens*
Dock, *Rumex spp.*
Duck potato, *Sagittaria spp.*
Duckweed, *Lemna spp.*

Eggplant, *Solanum melongena*
Elderberry, *Sambucus spp.*
Fennel, *Foeniculum vulgare*
Fenugreek, *Trigonella foenum-graecum*
Fig, *Ficus carica + spp.*
Filbert, *Corylus avellana + spp.*
Flax, *Linum spp.*
Florence fennel, *Foeniculum vulagare var. dulce*
Fuchsia, *Fuchsia spp.*
Garden cress, *Lepidium sativum*
Geranium, *Pelargonium spp.*
Ginger, *Zingiber officinale*
Gingseng, *Aralia quinquefolia*
Gladioli, *Gladiolus spp.*
Globe artichoke, *Cynara scolymus*
Glycine, *Neonotonia wightii*
Gliricidia, *Gliricidia sepium*
Gooseberry, *Ribes grossularia, R. uva-crispa*
Goosefoot, *Chenopodium album*
Gorse, *Ulex europaeus*
Granadilla, *Passiflora quadrangularis*
Grape, *Vitis vinifera*
Grass, banna, *Pennisetum purpureum*
  buffalo, *Stenotaphrum secundatum*
  elephant, *Pennisetum spp.*
  guinea, *Panicum maxium*
  Johnson, *Sorghum halespense*
  kikuyu, *Pennisetum clandestinum*
  lemon, *Cymbopogon citratus*
  pampas, *Cortaderia sellowiana*
  rice, *Oryzoides hymenopsis*
  vetiver, *Vetivaria zizanoides*
Guava, *Psidium guavaja*
Hawthorn, *Crataegus oxycanthus + spp.*
Hazelnut, *Corylus avellana + spp.*
Hickory, *Carya ovata*
Hops, *Humulus lupus*
Honey locust, *Gleditsia triacanthos*
Horseradish, *Armoracia rusticana*
Horseradish tree, *Moringa oleifera*
Huckleberry, *Gaylussacia, Vaccinium*
Ice cream bean tree, *Inga edulis*
Iceplant, *Mesembryanthemum spp.*
Indian water chestnut, *Trapa spp., T. natans*
Inga, *Inga spp., I. edulis*
Ivy, English, *Hedera helix,*
  Varigated, *H. corymbosa*
Jaboticaba, *Myrciaria cauliflora*
Jakfruit, *Artocarpus heterophyllus*

Jerusalem artichoke, *Helianthus tuberosus*
Jicama, *Pachyrrhizus erosus*
Jujube, *Ziziphus jujuba*
Kale, *Brassica oleracea var. acephala*
Kang kong, *Ipomoea aquatica*
Kiwifruit, *Actinidia chinensis*
Kiwifruit (hardy), *Actinidia arguta*
Kniphofia, *Kniphofia spp.*
Kurrajong, *Brachychiton populneum*
Lab-lab bean, *Lab-lab purpureus*
Lamb's quarters, *Chenopodium album*
Lantana, *Lantana camara*
Lavender, *Lavendula spp.*
Leatherwood, *Eucryphia billardierii*
Leeks, *Allium ampeloprasum*
Legumes, Fams: *Fagaceae, Vigna, Papilionaceae*
Lentils, *Lens culinaris*
Lespedeza, *Lespedeza spp.*
Leucaena, *Leucaena leucocephala*
Lettuce, *Latuca sativa*
Loquat, *Eriobotrya japonica*
Lotus, *Nelumbo nucifera*
Lovage, *Levisticum officinale*
Lucerne, *Medicago sativa*
Luffa gourd, *Luffa aegyptiaca*
Lupin, *Lupinus alba + spp.*
Lychee, *Litchi chinensis*
Marigold (African), *Tagetes erecta, T. minuta*
Macadamia, *Macadamia integrifolia*
Maize, *Zea mays*
Mango, *Mangifera indica*
Maple, *Acer saccharum*
Marsh tupelo, *Nyssa aquatica*
Mesquite, *Prosopis spp.*
Mexican blood trumpet, *Phaedranthus buccinatorius*
Millets, Fams: *Pennisetum, Panicum*
Mint, *Mentha spp.*
Mirror plant, *Coprosma repens*
Moringa, *Moringa oleifera*
Mulberry, *Morus spp.*
Mullein, *Verbascum thapsus*
Mustard, *Brassica nigra, B. hirta*
Nasturtium, *Tropaeolum majus*
Neem tree, *Azedarachta indica*
Nettle, *Urtica dioica*
Nutgrass, *Cyperus rotundus; Eleocharis*
Oak, *Quercus spp.*
  Chinquapin oak, *Q. prinoides*
  Cork oak, *Q. suber*

Black oak, *Q. velutina*
Durmast oak, *Q. petraea*
English oak, *Q. robur*
Holm oak, *Q. Ilex*
Pin oak, *Q. palustris*
Red oak, *Q. rubra*
White oak, *Q. alba*
Oats, *Avena sativa*
Okra, *Abelmoschus esculentus*
Olive, *Olea europea*
Onionweed, *Allium triquetrum*
Onions, *Allium spp.*
Osage orange, *Maclura pomifera*
Oxalis, *Oxalis spp.*
Oysternut, *Pelsairia occidentale*
Palm, butia, *Butia capitat borassus, Borassus flabellifer + spp.*
  Chilean wine, *Jubaea spectabilis*
  date, *Phoenix dactylifera*
  doum, *Hyphaene thebaicus*
  oil, *Elaeis guineaensis*
  peach, *Bactris Gasipaes*
Papaya, *Carica spp., C. papaya*
Parsley, *Petroselinum crispum*
Parsnip, *Pastinaca sativum*
Passionfruit, *Passiflora spp.*
Paulownia, *Paulownia spp.*
Paw paw, *Carica papaya*
Peach, *Amygdalus persicae*
Peanut, *Arachis hypogaea*
Pear, *Pyrus communis + spp.*
Peas, *Pisum spp., P. sativum*
Pecan, *Carya illinoensis*
Pennisetum, *Pennisetum spp.*
Pepino, *Solanum muricatum*
Pepper, chili, *Solanum frutescens*
  Sweet, *Solanum annuum*
Persimmon, *Diospyros kaki*
Pigeon pea, *Cajanus cajan*
Pigface, *Mesembryanthemum spp.*
Pine, Araucaria, *Araucaria spp.*
  Australian, *Callitris spp.*
  Cluster, *Pinus pinaster*
  Cuban, *Pinus caribaea*
  Norfolk Island, *Araucaria heterophylla*
  Pinyon, *Pinus edulis*
  Slash, *Pinus elliottii*
  Stone, *Pinus pinea*
Pineapple, *Ananus comosus*
Pineapple guava, *Feijoa sellowiana*
Pistachio, *Pistachia vera + spp.*
Plum, *Prunus domestica + spp.*
Pokeweed, *Phytolacca americana*
Pomegranate, *Punica granatum*
Pongamia, *Pongamia pinnata*
Poplar, *Populus spp.*

Potato, *Solanum tuberosum* pp.
Prickly pear cactus, *Opuntia spp.*
Pride of Madeira, *Echium fastuosum*
Prosopis, *Prosopis spp.*
Pultenea, *Pultenea spp.*
Pumpkin, *Cucurbita maxima*
Pussy willow, *Salix caprea*
Pyrethrum daisy, *Pyrethrum spp.*
Queen Anne's Lace, *Daucus carota var. carota*
Quinoa, *Chenopodium quinoa*
Rape, *Brassica napus*
Rambutan, *Alectryon subcinereus*
Raspberry, *Rubus idaeus + spp.*
Ragweed, *Ambrosia spp.*
Redwood, *Sequoia sempivirons*
Red-hot poker, *Kniphofia spp.*
Reed grass, *Phragmites spp.*
Rhubarb, *Rheum rhaponticum*
Rice, *Oryza sativa*
Rosella, *Hibiscus sabdariffa*
Rose, *Rosa multiflora*
Rose apple, *Eugenia malaccensis*
Rosemary, *Rosmarinus officinalis*
Rosewood (Burmese), *Pterocarpus indicus*
Round rushes, *Juncus effusos + spp.*
Rubber hedge (Africa), *Euphorbia tirucalli*
Russian olive, *Elaeagnus angustifolia*
Rye, *Secale cereale*
Safflower, *Carthamus tinctorius*
Sage, *Salvia officinalis*
Salt brushes, *Atriplex spp.*
Salsify, *Tragopogon porrifolius*
Salvia, *Salvia spp.*
Sapote, *Ciospyros, Casimiroa & genera*
Scarlet runner bean, *Phaseolus coccineus, P. multiflorus*
Seagrasses, *Posidonia, Zostera spp.*
Sedge, *Scirpus spp. S. validus, Cyperus spp.*
Serviceberry, *Amelanchier canadensis*
Sesame, *Sesamum indicum*
Sesbania, *Sesbania spp.*
Shepherd's purse, *Capsella bursa-pastoris*
Shallot, *Allium aggregatum* group
Siberian pea shrub, *Caragana spp.*
Shiitake (fungus), *Lentinus almum*
Silky oak, *Grevillea robusta*
Silverbeet, *Beta oleracea var. acephala*

Silverberry, *Elaeagnus commutata*
Sodom apple, *Solanum spp.*
Southernwood, *Artemesia abrotanum*
Spinach, *Spinacia oleracea*
Strawberry, *Fragaria vesca + spp.*
Sugar beet, *Beta vulgaris*
Sugar cane, *Saccarum officinarum*
Sunflower, *Helianthus annuus*
Sunn hemp, *Crotalaria juncea*
Sunroot (Jerusalem artichoke), *Helianthus tuberosus*
Swamp holly, *Ilex, Amelanchier*
Swede, *Brassica napus var. napobrassica*
Sweet cicily, *Myrrhis odorata*
Sweetgrass, *Glyceria*
Sweet potato, *Ipomoea batatus*
Sweet woodbine, *Louicera caprifolium*
Swiss chard, *Beta oleracea var. acepahal*
Tagasaste, *Chaemocytisus palmensis*
Tamarillo, *Cyphomandra betacea*
Tamarind, *Tamarindus indicus*
Tamarisk, *Tamarix apetala & spp.*
Tansy, *Tanacetum vulgare*
Tapioca, *Manihot esculenta*
Taro, *Colocasia esculenta*
Tarragon, *Artemesia dracunculus*
Taupata, *Coprosma repens*
Tea, *Camellia sinensis*
Teak, *Tectona grandis*
Tea tree, *Leptospermum, Melaleuca spp.*
Thistle, *Cnicus benedictus*
Thyme, *Thymus spp., T. vulgaris*
Tipuana tipu, *Tipuana tipu*
Tobacco, *Nicotiana tabacum*
Tobacco bush, *Nicotiana spp.*
Tomatillo, *Physalis ixocarpa*
Tomato, *Lycopersicon lycopersicum*
Turmeric, *Curcuma domestica*
Turnip, *Brassica rapa var. septiceps*
Vanilla, *Vanilla planifolia*
Vetch, *Vicia spp.*
Virginia creeper, *Parthenocissus quinquefolia*
Walnut, *Juglans regia*
Wandering jew, *Tradescantia albiflora*
Water chestnuts, *Eleocharis, Trapa spp.*
Water elm, *Ulmus aquatica*
Waterlily, *Nymphaea spp.*
Water mint, *Mentha aquatica*

Watercress, *Rorippa amphibia + spp.*
Watermelon, *Citrullus vulgaris*
Water ribbon, *Triglochin*
Wattle, green, *Acacia mearnsii* silver, *Acacia dealbata*
Wheat, *Triticum spp., T. aestivum*
White acacia, *Acacia albida*
White cedar, *Melia azedarach*
White clover, *Trifolium repens*
Wild rice, *Zizania lacustris*
Willow, *Salix spp.*
Wisteria, *Wisteria floribunda*
Wormwood, *Artemesia absynthium*
Yam, *Dioscorus spp.*
Yarrow, *Achillea millefolium*
Yatay, *Butia capitata, B. yatay*
Yew, *Taxus spp.*
Youngberry, *Rubus ursinus*

B: PLANTS MENTIONED IN TEXT BY SPECIES NAME

Abelmoschus esculentus, Okra
Acacia spp., Acacias
  A. acuminata, Raspberry jam acacia
  A. albida, White acacia
  A. aneura, Mulga
  A. armata, Kangaroo thorn
  A. auriculiformis, *A. mangium*, Tropical acacias
  A. baileyana, Cootamundra wattle
  A. dealbata, Silver wattle
  A. fimbriata, Brisbane wattle
  A. mearnsii, Green wattle
  A. melanoxylon, Blackwood, sally wattle
  A. saligna, Weeping wattle
  A. sophorae, *A. longifolia*, Golden wattle
Actinidia arguta, Hardy kiwifruit
  A. chinensis, Kiwifruit, Chinese gooseberry
Agave spp., Agave
Albizia spp. Albizia
  A. lopantha, Coast albizia
Allium spp., Onion group
Aloe spp., Aloe
Alyssum spp., Alyssum
Amaranthus spp., Amaranth
Amelanchier canadensis, Serviceberry
Amygdalus persicae, Peach
Anacardium occidentale, Cashew
Ananus comosus, Pineapple
Anethum graveolens, Dill
Annona spp., Custard apple
Apium graveolens, Celery
Arachis hypogaea, Peanut
Aralia quinquefolia, Gingseng
Araucaria spp., Araucaria pine
  A. heterophylla, Norfolk Island pine
Arctotheca calendula, Capeweed
Armeniaca vulgaris, Apricot
Armoracia rusticana, Horseradish
Artemesia absinthium, Wormwood
Artocarpus spp., Jakfruit
  A. altilis, Breadfruit
Arundinaria spp., Bamboo
Asparagus officinalis, Asparagus
Aster spp., Aster, daisies
Atriplex spp., Salt bushes
Avena sativa, Oats
Azedarachta indica, Neem tree

Azolla spp., *A. filicoides*, Azolla
Bambusa spp., Bamboo
Banksia spp., Banksia
Beta vulgaris, Beets, sugar beet, silver beet, swiss chard
Betula spp., Birch
Borassus flabellifer, Borassus palm
Brachychiton australis, Bottle tree
Brassica napus, Rape
  B. nigra, *B. hirta*, Mustard
  B. oleracea, Broccoli, cauliflower, kale
  B. rapa, Turnip
Butia capitata, Butia palm, jelly palm
  B. yatay, Yatay
Cajanus cajan, Pigeon pea
Calocarpum spp., Sapote
Calliandra spp., Calliandra
Callitris spp., cypress pine
Camellia sinensis, Tea
Canna edulis, Queensland arrowroot
Capsicum annuum, Capsicum, bell pepper
Caragana arborescens, Siberian pea shrub
Carica spp., *C. papaya*, Papaya, paw paw
Carthamus tinctorius, Safflower
Carya ovata, Hickory
Casimiroa spp., Sapote
Cassia spp., Cassia
Castanea spp., Chestnut
  C. pumila, Chinquapin
Casuarina spp., Casuarina, she-oak
Cedrus spp., Cedar
Celosia, Woolflower
Ceratonia siliqua, Carob
Chaemocytisus palmensis, Tagasaste
Chenopodium spp., Goosefoot, fat hen
  C. quinoa, Quinoa
Cichorium intybus, Chicory
Citrullus vulgaris, Watermelon
Citrus spp., Citrus
Cocos nucifera, Coconut
Coffea spp., Coffee
Colocasia esculenta, Taro
Coprosma repens, Taupata, mirror plant
Cordia abyssinica, Cordia
Cortaderia sellowiana, Pampas grass
Corylus avellana + spp., Hazelnut, filbert
Crataegus oxycanthus + spp., Hawthorn
Crocus sativus, Crocus

Crotalaria spp., Crotalaria, rattle pod, sunn hemp
Cucumis sativus, Cucumber
  C. melo, Melons
Cucurbita maxima, Pumpkin
Curcuma domestica, Turmeric
Cydonia oblonga, Quince
Cymbopogon citratus, Lemongrass
Cynara scolymus, Globe artichoke
Cyperus rotundus, nutgrass
Daucus carota, Carrot
Derris spp., *D. elliptica*, Derris
Dichondra repens, Dichondra
Digitaria decumbens, Pangola
  D. exilis, Couch grass
Dioscorus spp., Yam
Diospyros spp. Sapote
  D. kaki, Persimmon
Echium fastuosum, Pride of Madeira
Elaeagnus angustifolia, Russian olive
  E. umbellata, Autumn olive
Elaeis guineaensis, Oil palm
Eleocharis sphacelata + spp., Nutgrass
  E. dulcis, Chinese water chestnut
Eriobotrya japonica, Loquat
Eucalyptus spp., Eucalypts
Eucryphia billardierii, Leatherwood
Euphorbia tirucalli, Rubber hedge (Africa)
Fagopyrum esculentum, Buckwheat
Ficus carica + spp., Fig
Feijoa sellowiana, Feijoa, pineapple guava
Foeniculum vulgare, Fennel
Fragaris vesca + spp., Strawberry
Fuchsia spp., Fuchsia
Galium aparine, Cleavers
Gaylussacia spp., Huckleberry
Gladiolus spp., Gladioli
Gleditsia triacanthos, Honey locust
Gliricidia sepium, Gliricidia
Glycine max, soya bean
Gmelina spp., Gmelina
Gossypium spp., Cotton
Grevillea robusta, Silky oak
Helianthus annuus, Sunflower
  H. tuberosus, Sunroot, Jerusalem artichoke
Hibiscus sabdariffa, Rosella
Hordeum vulgare, Barley
Hyphaene thebaicus, Doum palm
Inga edulis, Inga, ice-cream bean tree
Ipomoea aquatica, kang kong
  I. batatas, sweet potato
Juglans nigra, Black walnut
  J. regia, Walnut

Juncus effusos + spp., Round rushes
Kniphofia spp., Kniphofia, red-hot poker
Lab-lab purpureus, Dolichos bean, lab-lab bean
Lantana camara, Lantana
Latuca sativa, Lettuce
Lavendula spp., Lavender
Lens culinaris, Lentils
Lemna spp., Duckweed
Lentinus edodes, Shiitake mushroom
Leucaena leucocephala, Leucaena
Litchi chinensis, Lychee
Lupinus alba + spp., Lupin
Lycium ferrocissimum +spp., African boxthorn
Lycopersicon lycopersicum, Tomato
Macadamia integrifolia, Macadamia nut
Malus pumila, Apple
Mangifera indica, Mango
Manihot esculenta, Cassava, tapioca
Maranta arundinaceae, Arrowroot (West Indian)
Medicago sativa, Lucerne, alfalfa
Melia azedarach, Melia, white cedar
Mentha aquatica, Water mint
Mesembryanthemum spp., Pigface, iceplant
Monstera deliciosa, Ceriman
Moringa oleifera, Horseradish tree
Morus spp., Mulberry
Musa paradisiaca + spp., Banana
Narcissus, Daffodil
Nasturtium var., Nasturtium
Nelumbo nucifera, Lotus
Nepeta cataria, Catmint
Nicotiana spp., Tobacco, tobacco bush
N. alata, Flowering tobacco
Aruacaria heterophylla, Norfolk Island pine
Nymphaea spp. Waterlily
Ocimum basilicum, Basil
Olea europea, Olive
Oryza sativa, Rice
Oxalis spp., Oxalis
Pachyrrhizos tuberosus, Yam bean, jicama
Panicum spp., Millets, panic grass
Pastinaca sativum, Parsnip
Passiflora spp., Passionfruits
Pennisetum spp., Pennisetum, elephant grass

Persea americana, Avocado
Phaseolus spp., Beans
Phoenix dactylifera, Date plam
Physalis peruviana, Cape gooseberry
Pinus spp., Pines
P. caribaea, Cuban pine
P. elliottii, Slash pine
P. pinaster, Cluster pine
P. coulteri, Big cone pine
P. edulis, Pinyon pine
P. pinea, Stone pine
Pistachia vera + spp., Pistachio
Pisum spp., Peas
Pongamia pinnata, Pongamia
Populus spp., Poplar
Prosopis spp., Prosopis, mesquite
Prunus spp., Cherry, plum, almond
Psidium guavaja, Guava
Psophocarpus tetragonolobus, Winged bean
Pultenea spp., Pultenea
Punica granatum, Pomegranate
Pyrethrum spp., *P. cinerariifolium*, Pyrethrum daisy
Pyrus communis + spp., Pear
Quercus spp., Oaks
Raphanus sativus, Daikon radish
Rheum rhaponticum, Rhubarb
Ribes spp., Currants, gooseberries
Robinia pseudoacacia, Black locust
Rorippa amphibia + spp., Watercress
Rosa multiflora, Rose (hedgerow)
Rosmarinus officinalis, Rosemary
Rubus spp., Blackberry, raspberry, loganberry, etc.
Rumex spp., Dock
Ruppia martina, Watergrass
Saccharum officinarum, Sugar cane
Sagittaria spp., Arrowhead, duck potato
Salix spp., Willow
Salvia spp., Salvia, sage
Sambucus spp., Elderberry
Scirpus spp., *S. validus*, Sedge
Secale cereale, Rye
Sechium edule, Chayote, choko
Sesbania spp., Sesbania
Sequoia sempivirens, Redwood
Setaria spp., Millets, grass
Solanum spp., Eggplant, sodom apple, pepino, nightshade
Sorghum almum, silk sorghum
S. halapense, Sudan grass
Spinacia oleracea, Spinach

Symphytum officinale, Comfrey
Tagetes erecta, T. minuta, African marigold
Tamarindus indicus, Tamarind
Tamarix apetala + spp., Tamarisk
Tarascum officinale, Dandelion
Tectona grandis, Teak
Theobroma cacao, Cocoa
Thymus spp., T. vulgaris, Thyme
Tipuana tipu, Tipuana tipu, Pride of Bolivia
Tradescantia albiflora, Wandering jew
Trapa natans, Indian water chestnut
Trifolium spp., Clover
Trigonella foenum-graecum, Fenugreek
Triticum aestivum, Wheat
Typha spp., Cattail, cumbungi
Ulex europaeus, Gorse
Urtica dioica, Nettle
Vaccinium spp., Blueberry, huckleberry
V. macrocarpon + spp., Cranberry
Vanilla planifolia, Vanilla
Vetiveria zizanoides, Vetiver grass
Vicia faba, Fava bean, broad bean
Vigna spp., Beans
Vitis vinifera, Grape
Zea mays, Corn, maize
Zingiber officinale, Ginger
Ziziphus jujuba, Jujube
Zostera spp., Seagrass

Appendix D

GLOSSARY

Aspect: View that is facing a certain direction, e.g. a sunny aspect faces the sun. An eastern aspect receives sun in the morning, whereas a western aspect receives afternoon sun.

Afforestation: Planting trees in an area where logging has occurred or where trees have not previously grown.

Allelopathy: The process by which plants release toxins through their leaves or roots to inhibit the growth of other, nearby plants. Some examples are black walnut, sunflower, sunroot, and barley.

Annual: A plant that completes its life cycle in a single growing season, setting seed and dying. Annuals take more energy (work, power) so should be located close to the home.

Aquaculture: Raising and managing fish, other aquatic organisms, and plants in specially-prepared ponds rather than harvesting from the wild.

Biennial: A plant which flowers, sets seed, and dies in its second year. Examples of biennials include pineapple, leeks, kale, and fennel. Many biennials are grown as annuals in cold climates.

Coppice: Cutting of trees or shrubs which then resprout via branches or root suckers. Examples of such trees are willows, eucalypts, alder, and leucaena.

Cover crops: Plants grown to protect soil from erosion and to provide organic material. Cover crops are usually grown in young orchards, gardens, and cropland during the cold season, and are sometimes turned under before flowering and setting seed (see green manuring).

Deciduous: Plant which loses its leaves in winter, useful when planting near the house so that summer sun can shine through the bare branches.

Dioecious: Plants which bear male and female flowers on separate plants; both are needed for pollination and fruit set, usually with a ratio of 1 male to 5-20 females.

Edge: The junction/zone that lies between two media or landscape forms; a border where materials or resources accumulate.

Espalier: A tree or shrub which is trained against a wall, fence, or trellis. Useful mostly for small gardens as it requires time and energy.

Guild: A species assembly of plants and animals which benefit each other, usually for pest control.

Green manure: Plants which are turned into the soil to enhance fertility; these are mostly legumes.

Intercropping: A system of growing two or more crops side by side or intermixed on the same area of ground.

Keyline: System of water conservation developed by P.A. Yeomans using underground channels to recharge groundwater supplies. Series of well-placed dams are also integral to the system.

Legumes: Plants of the family Leguminoseae (e.g. beans, peas, clovers, and tree legumes such as acacia, albizia, and cassia). Most legumes (but not all e.g. honey locust, carob) fix atmospheric nitrogen in the soil through a symbiotic relationship with a bacteria within their roots. This nitrogen is available to the plant, but not necessarily to other plants unless coppiced or turned into the soil.

Microclimate: The localised climate around landscape features and buildings; important for selecting sites for specific crops or species.

Monoculture: A crop of plants of the same kind on a piece of land, usually giving rise to severe pest infestation.

Multi-storey: A mixture of plant species comprising a ground layer, shrubs, and trees of varying heights.

No tillage: No cultivation of the soil, using instead a combination of tree crop, mulch, and green manure to build soil fertility. Weeds are controlled by slashing, mulching, browsing, or flooding.

Nurse plants: Pioneer species used to provide green manure, nutrient, or shade for succeeding crops or trees.

Perennial: Plant which lives longer than one or two years, and which usually sets flowers/fruit every year (after a certain age is reached).

Pioneers: Plant species which populate vacant ground and which eventually give way to other species.

Predator species: Insects or vertebrates eating pest species, e.g. ladybird larvae controlling aphids.

Polyculture: The planting of multiple crops in the same ground or area.

Rhizobia: Bacteria which form nodules in the roots of most legumes and fix atmospheric nitrogen in the soil.

Solar chimney: A black, thin metal chimney which acts as a heat engine to exhaust air from a room or enclosed place, thus drawing in fresh or cool air.

Stacking: Arrangement of plants to take advantage of all possible space, using tall and medium-sized trees with a lower shrub and herb layer. Care must be taken so that water and light competition are at a minimum.

Succession: Progressive change from one plant (and animal) community to another. Permaculture seeks to accelerate succession by using and managing pioneer species (weeds) rather than setting the system back by weeding.

Swales: Long, level excavations made to intercept and hold runoff water. The water slowly infiltrates into the ground, benefiting trees and shrubs planted on the downhill bank.

Thermal belt: Sun-facing mid-slope site defined by fewer frosts (hence earlier leaf and bud formation); a good site for homes and crops.

Wildlife corridors: Belts of trees, marsh, or river forest connecting two or more larger habitat areas.

Appendix E

PERMACULTURE MAGAZINES & NEWSLETTERS

Subscriptions to the following quarterly mazagines will give you information on
permaculture themes, news and upcoming events, training programmes,
contact addresses, and other useful data.

International Permaculture Journal, PO Box 6039, Sth Lismore NSW 2480, Australia. $15/yr (overseas seamail $20). Ph: (066) 220020. New Zealand subscribers: $29 to Permaculture N.Z., PO Box 37030, Parnell, N.Z.). Quarterly magazine contains articles, resources listings, book reviews. Back issues highly recommended.

The Permaculture Edge, Permaculture Nambour, PO Box 650, Nambour QLD 4560, Australia. $16/yr; quarterly magazine devoted to worldwide practical application of permaculture.

The Permaculture Activist, Route 1, Box 38, Primm Springs, TN 38476, USA. Ph: (615) 583 2294. $13/yr. A quarterly journal for the permaculture movement of North America; also U.S. agents for the *International Permaculture Journal*. (US$20/yr).

The Drylands Journal, Permaculture Drylands Institute, PO Box 133, Pearce, AZ 85625 USA. Ph: (602) 824 3465. Excellent articles on plants and strategies for dryland environments.

Permanent Publications, Hyden House Limited, Little Hyden Lane, Clanfield, Hants PO8 ORU U.K. Ph: (0705) 596500. Quarterly magazine with news and articles from British permaculture designers/teachers. £2 per issue.

Permaculture Africa Newsletter, The Botswana Permaculture Trust, Private Bag 47, Serowe, Botswana. Ph: 430550/430930.

PC Association of Zimbabwe Newsletter, PO Box 8515, Causeway, Harare, Zimbabwe.

There are many other permaculture newsletters, associations, and groups in Australia and all over the world.
For a full directory, send a stamped, self-addressed envelope to the *International Permaculture Journal*.
(Australia only—all others may send international postal stamps totalling AUD $1.20).

PERMACULTURE GROUPS IN AUSTRALIA

Permaculture Sydney Association
33 Bundock Street
Randwick NSW 2031
Ph: (02) 344 0956

Permaculture Melbourne
6 Derby St
Kew Vic 3101
Ph: (03) 853 6828

Permaculture South Australia
1st flr, 155 Pirie St
Adelaide SA 5000
Ph: (08) 379 7492

Permaculture Association of W.A.
PO Box 430
Subiaco WA 6008
Ph: (09) 417 2274

Permaculture A.C.T.
PO Box 886
Civic Square ACT 2608
Ph: (06) 248 5192

The Web
142 Agnew St
Norman Park QLD 4170
Ph: (07) 899 1953

SOME USEFUL RESOURCES

Rainbow Power Company Pty Ltd
70 Cullen St
Nimbin NSW 2480
Ph: (066) 89 1430
Fax: (066) 89 1109

Rainbow manufactures, sells and installs a wide range of solar, hydro, and wind energy systems.
Write for catalogue enclosing 2 stamps.

Seed Savers Network
PO Box 975
Byron Bay NSW 2481
Ph: (066) 856 624

Collects, catalogues and maintains open-pollinated, non-hybrid seeds, specialising in traditional vegetable varieties.
Twice yearly newsletter. SAE for information.

Eden Seeds
MS316
Gympie QLD 4570

Suppliers of non-hybrid vegetable, tree, legume and herb seeds. Send 2 stamps for catalogue.

Pegasus Network
PO Box 248
Broadway QLD 4006
Ph: (07) 257 1111
Fax: (07) 257 1087

A computer message and conferencing system for peace, human rights and the environment. Access for a local call fee from anywhere in Australia to electronic mail boxes.

Seeds of Change
1364 Rufina Circle 15
Santa Fe NM 87501
USA
Ph: (505) 438 8080
Fax: (505) 438 7052

Suppliers of a wide range of organic seeds. Phone for catalogue.

Biographies

BILL MOLLISON was born in 1928 in the small fishing village of Stanley, Tasmania, and left school at fifteen to help run the family bakery. He soon went to sea as a shark fisherman, and until 1954 filled a variety of jobs as a forester, mill-worker, snarer, and naturalist.

Bill joined the CSIRO (Wildlife Survey Section) in 1954 and for the next nine years worked in many remote locations in Australia as a biologist. In 1963 he spent a year at the Tasmania Museum in curatorial duties, then returned to field work with the Inland Fisheries Commission.

Returning to formal studies in 1966, he lived on his wits running cattle, bouncing at dances, shark fishing, and teaching part-time at an exclusive girls' school. Upon receiving his degree in bio-geography, he was appointed to the University of Tasmania. In 1974, he and David Holmgren, then a student at the University, developed and refined the permaculture concept leading to the publication of *Permaculture One* and *Permaculture Two*.

Since leaving the university in 1978, Bill has devoted all his energies to furthering the system of permaculture and spreading the idea and principles worldwide. He has taught thousands of students, and has contributed many articles, curricula, reports, and recommendations for farm projects, urban clusters, and local government bodies. In 1981, Bill received the Right Livelihood Award in Stockholm for his work in environmental design. He has recently received the inaugural Vavilov medal for his significant contribution to agricultural science (USSR) and in 1993 was honoured as an Outstanding Australian Achiever.

Bill Mollison is the executive director of the Permaculture Institute, which was established in 1979 to teach the practical design of sustainable soil, water, plant, and legal and economic systems to students worldwide. He is the author of *Permaculture - A Designers' Manual* (1988), *Ferment and Human Nutrition* (1993) and the father of six children.

RENY MIA SLAY grew up in the Canary Islands, where her father was a teacher and market gardener. Returning to the USA for university studies, she was caught up in the back-to-the-land movement of the early 1970s, co-authoring a book (*Homesteaders' Handbook*) and working for a summer on one of California's first organic farms.

After a job in Mexico, came three years at the Farallones Institute's Rural Centre as office, manager, workshop organiser, tour guide, and edible landscaping apprentice. Drawn to permaculture, Reny moved to Tasmania to become Bill Mollison's 'expedition leader', arranging teaching trips to Europe, New Zealand, USA, Nepal and outback Australia. As a manager of Tagari Publications until 1988, she worked closely with Bill, editing the latest permaculture books.

Reny, a talented artist, now lives in the lush rainforest valley of the ancient Wollumbin Caldera of New South Wales.

THE PERMACULTURE INSTITUTE

In 1979 The Permaculture Institute was established to teach the practical design of sustainable land use and community development to farmers, horticulturalists, community aid workers, business people, landscape designers, and individuals interested in ethical land use.

An important focus of the Permaculture Institute is on education. What is taught is a way of thinking, not just on how to re-establish an eroded area or plan an integrated chicken forage/ orchard area. Permaculture represents an educational process that promotes responsible thinking and stresses positive ethics.

Although firmly based around natural gardening and farming methods, permaculture is a much broader, more wholistic design system. It shows not just how to produce food in a way which doesn't cost the earth, but how to achieve sustainable human settlements, preserve and extend natural ecosystems, and live gently on the earth in all aspects of our lives.

Since 1981, when the first permaculture design course was held, thousand of people have been trained worldwide. Many more have attended permaculture lectures, weekend workshops, conferences, and field days. There are local permaculture associations, consultancies and groups in Australia, U.K., Mexico, South America, Canada, New Zealand, Europe, and Africa, many with local newsletters which inform about regional needs. These groups are linked together by a variety of journals containing informative articles, directories of events, courses, and book reviews.

The permaculture organization, as a whole, is run by small groups which are completely independent. Everyone does what they can, either on a small backyard scale, or in the larger community arena. Every two years, a Permaculture Conference is held in a different country. Design course graduates share their skills and experience at these conferences, discuss and show slides of their work, policy is formulated and new directions are discussed.

Permaculture has had a profound influence on people all over the world, from ordinary farmers and suburban homeowners to government officials and teachers, in inspiring people to work towards positive action of earth care. There are no grand projects or expensive showcase sites, rather than the sum total of many people working in their backyards, farms, villages, and local communities.

"All of us acknowledge our own work is modest; it is the totality of many modest works that is impressive." (B. Mollison)

THE PERMACULTURE INSTITUTE, TYALGUM

The Permaculture Institute is based in sub-tropical northern New South Wales, where Bill Mollison has designed and implemented a permaculture design. The land, bought in mid 1987, was formerly grazing land, and had been overgrazed and burnt. Only three trees remained of the once-extensive forest, still visible on the surrounding mountainsides.

Five years later, hundreds of trees have been planted, carefully placed to provide shelter, mulch, wildlife habitat, bee forage, fruit, nuts and fuelwood. The site is planned according to permaculture principles, with multi-function plants, dams and ponds for aquaculture and wetland flora and fauna, a home garden and orchard, woodlot and small rainforest area.

The two-storey brick house acts as a home for Bill, an office for Tagari Publications (publishers of the permaculture books), and as a central information base for all the Permaculture associations and groups worldwide.

Throughout the year, visitors walk the property, some of them school children, organized by local teachers.

The Institute has become a focus for local activities, providing a database facility, and acting as a bioregional office. Local people have shown great interest and enthusiasm for the methods used here, and many have themselves attended permaculture courses.

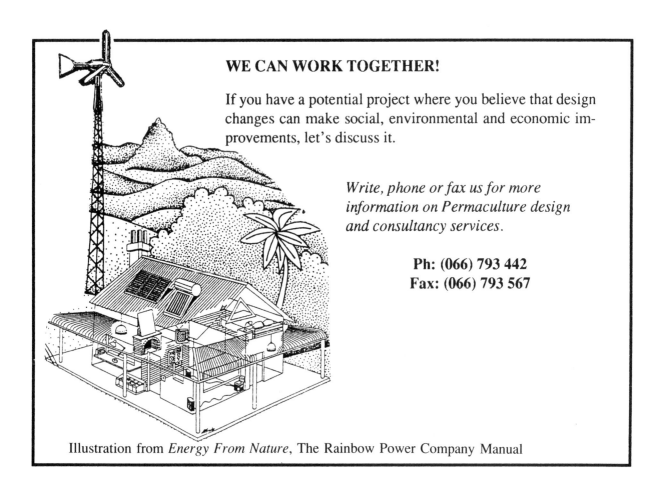

WE CAN WORK TOGETHER!

If you have a potential project where you believe that design changes can make social, environmental and economic improvements, let's discuss it.

Write, phone or fax us for more information on Permaculture design and consultancy services.

Ph: (066) 793 442
Fax: (066) 793 567

Illustration from *Energy From Nature*, The Rainbow Power Company Manual

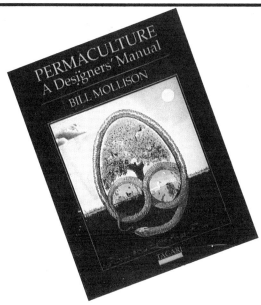

PERMACULTURE
A Designers' Manual

by
Bill Mollison

**This is the definitive permaculture design manual covering
all aspects of property design and natural farming techniques.**

- Microclimate and broadscale techniques
- Species selection, placement and management
- Multi-purpose shelterbelt, woodlot, orchard and forage systems
- Plant succession and ecology
- Revegetation and afforestation techniques
- Home gardens
- Arid and humid landscape techniques and strategies
- Soil conservation and rehabilitation of eroded lands
- Water harvesting and irrigation systems
- Earthworks - terraces, swales, dams and canals
- Ecological principles and practices
- Forest systems
- Wildlife management
- Recycling and waste disposal
- Bioregional organisation and land access strategies
- Community financing systems
- Business strategies
- Ethical values for a new world
- And much, much more!

$65.00
(inc. postage)

580 pages   •   130 colour photos   •   450 line drawings

ISBN 0 908228 01 5

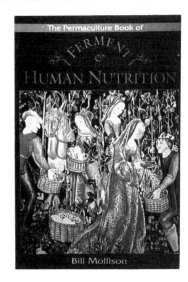

The following pages are excerpts from two Permaculture designs; the first is extracted from a report prepared by Bill Mollison and Birgit Seidlich for a sub-tropical farm.

The second design, by Robyn Frances, shows how a well-designed suburban block can supply large quantities of fresh fruit & vegetables year-round.

This report contains an analysis and a permaculture design concept for the property of E & J Brown, Rosebank, NSW. The analysis is based on an inventory of the natural features of the property such as landform, water, existing vegetation, soil structure, access and built structures.......

The design concept provides guidelines and general construction schedules for the implementation of organic agricultural systems, a nursery, a tearoom and seminar building, the residence and home garden. We have also provided a resource list with references and contact details for additional information on marketing, and the construction of buildings, energy systems and dams. A plant list recommends useful plant species with a description of their individual needs and uses.......

OBJECTIVES
Following consultation with the client, we have agreed on a set of objectives for the development of the property.
- self-sustainable agricultural systems
- maximum yield on minimum area
- low labour intensity
- potential for social income via services or workshops
- energy self-sufficiency for residential living
- minimal soil disturbance and erosion prevention
- minimal water run-off and optimum utilisation of on-site water resources.................

MANAGEMENT ZONES
We identified eight management zones on the property. Each of these zones will have a primary function suited to its natural characteristics and production potential. The zones will require individual management to ensure their proper maintenance. They are designed to sup-

port each other to develop into ecologically balanced systems. The eight zones are defined as:

1   Residence and home garden.
2   Nursery
3   Teahouse and seminar room
4   Orchard
5   Agricultural land
6   Riverine forest and recreation area
7   Ridgeline road
8   Camping

The zones are marked on Map 1 in the appendix.

The general layout and positioning of structures will ideally achieve a light net of interrelating elements.

We want to avoid wasting human energy equally as much as the unnecessary use of fuels and electricity, thus places frequently visited are grouped together.

The residence, the nursery and the tearoom form the central core of the property and here people will spend most of the day working or at leisure, and therefore most resources should be located near them, such as water supply, tools, workshop, animal shelters, and of course food.

It is very time-consuming to walk 15 minutes uphill every day to pick corn for dinner and would be an unpleasant task in bad weather. A 10 minute walk to pick lemons from the orchard adds up to 20 minutes a day because the fruit has to be brought back to the house. If we pick lemons twice a week for one year, the time adds up to 36 hours = 1.5 days work, just to be able to squeeze lemon juice onto our salads. With the lemon tree situated in the kitchen garden, a few steps away from the door, 36 hours = 1.5 days can be saved towards the annual holiday.

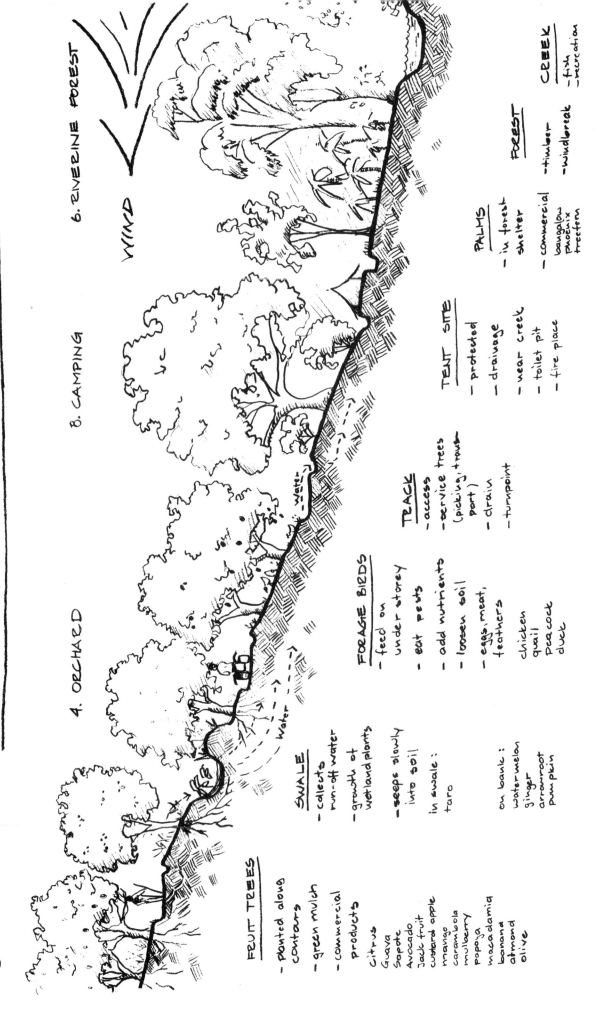

SECTION OF CONCEPT FOR EAST SLOPE

6. RIVERINE FOREST

WIND

8. CAMPING

4. ORCHARD

FRUIT TREES
- planted along contours
- green mulch
- commercial products

Citrus
Guava
Sapote
Avocado
Jack fruit
custard apple
mango
carambola
mulberry
papaya
macadamia
banana
almond
olive

SWALE
- collects run-off water
- growth of wetland plants
- seeps slowly into soil

in swale: taro

on bank: watermelon ginger arrowroot pumpkin

FORAGE BIRDS
- feed on under-storey
- eat pests
- add nutrients
- loosen soil
- eggs, meat, feathers

chicken
quail
peacock
duck

TRACK
- access
- service trees (picking, transport)
- drain
- turnpoint

TENT SITE
- protected
- drainage
- near creek
- toilet pit
- fire place

PALMS
- in forest shelter
- commercial bangalow phoenix treefern

FOREST
- timber
- windbreak

CREEK
- fish
- recreation

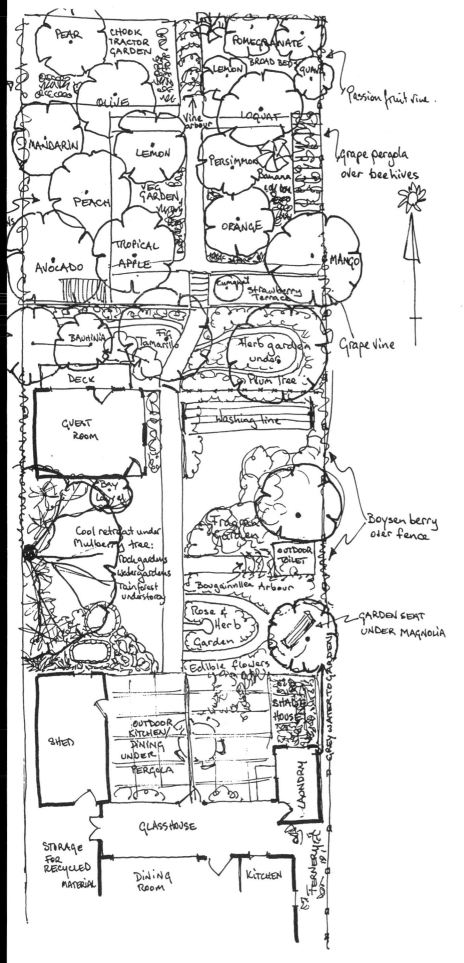

PEAR
CHOOK TRACTOR GARDEN
POMEGRANATE
LEMON
BROAD BEDS
GUAVA
OLIVE
Vine arbour
LOQUAT
Passion fruit vine.
MANDARIN
LEMON
PERSIMMON
Banana
Grape pergola over beehives
VEG GARDEN
PEACH
ORANGE
TROPICAL APPLE
AVOCADO
MANGO
Kumquat
Strawberry Terrace
BAUHINIA
Fig
Tamarillo
Herb garden under
Grape vine
DECK
Plum Tree
GUEST ROOM
Washing line
Bay Leaf
Fragrant garden
Cool retreat under Mulberry tree:
rock gardens
water gardens
rainforest understorey
Boysen berry over fence
OUTDOOR TOILET
Bougainvillea Arbour
Rose & Herb Garden
GARDEN SEAT UNDER MAGNOLIA
Edible flowers
SHED
SHADE HOUSE
OUTDOOR KITCHEN/ DINING UNDER PERGOLA
LAUNDRY
GLASSHOUSE
R. GREY WATER TO GARDEN
FERNERY/
STORAGE FOR RECYCLED MATERIAL
DINING ROOM
KITCHEN

Sample design for a long narrow ¼ acre block in Sydney's inner western suburbs. An outdoor living area merges houses & garden. A food forest in the lower backyard includes chickens & bees.

Design created for Frank & Laurel Pink by Robyn Francis, Permaculture Education, PO Box 379, Nimbin NSW 2480.

INDEX

NOTES

NOTES

NOTES